Math Mammoth
Subtraction 1

By Maria Miller

Contents

Introduction

Math Mammoth Subtraction 1 is a worktext that covers the concept of subtraction, the relationship between addition and subtraction, and the various meanings of subtraction.

In the first lesson, *Subtraction Is Taking Away*, the child learns the basic meaning of subtraction as taking away objects, and learns to write subtractions from an illustration where some objects are crossed out. The child can figure out the answers by simply counting how many objects are left.

If the child does not yet know the word "minus," it is a good idea to introduce it first orally. Use blocks or other concrete objects. For example, show the child eight blocks and take away three blocks. Then use both kinds of wordings: "Eight blocks, take away three blocks, leaves five blocks. Eight blocks *minus* three blocks *equals* five blocks." Then let the child do the same. Play with concrete objects until the child can use the words "minus" and "equals" in their own speech.

In the next lesson, the child counts down to subtract, which ties in subtraction with the number line. This is a transitional strategy to solve subtraction problems, because later students will learn more efficient ways to subtract, but it is important conceptually. For now, the student can solve $9 - 3$ by counting down three steps from nine: eight, seven, six. So the answer is six.

The following lesson, *Subtraction and Addition in the Same Picture*, begins the study of the relationship between addition and subtraction. This concept will span several lessons. This first lesson presents two sets of objects, such as blue and white balls, and the student writes both an addition sentence and a subtraction sentence from this illustration.

The lesson *When Can You Subtract?* concentrates on the idea that some subtractions, such as $4 - 5$, are meaningless when you think of taking away. The child also makes subtraction patterns in this lesson.

Then we continue studying the connection between addition and subtraction in the lesson *Two Subtractions from One Addition*. As an example, the child writes both $8 - 3 = 5$ and $8 - 5 = 3$ from the addition $3 + 5 = 8$. This idea ties in with fact families, a concept that is coming up soon.

In the lesson *Two Parts—One Total,* we study word problems that do not involve the idea of taking away but have two parts making up a total. For example, if there are 10 flowers of which some are white and some are red, and seven of them are white, how many are red? We know the "parts" (the red and white flowers) add up to 10, so we can write a missing addend sentence $7 + \_\_ = 10$. This can be solved by subtracting $10 - 7$ or by knowing the addition fact $7 + 3 = 10$.

Then we study fact families. This means writing two additions and two subtractions using the same three numbers.

In the lesson *How Many More?* students find how many more or how many fewer objects one person has than the other by drawing the objects. This lesson can easily be done with manipulatives if desired.

In the very next lesson, *"How Many More" Problems and Differences,* we continue the theme, this time writing a missing addend addition for problems that ask "how many more." For example, Veronica has 4 marbles and Ann has 6. We write the missing addend sentence $4 + \_\_\_ = 6$ to find how how many more Ann has. In the next lesson the child then learns to write subtraction sentences for such problems.

Besides the written problems, I encourage you to use the following games. Children like to play, and using the addition and subtraction facts in a game gives them fun and education in the same "package".

I hope you find this book helpful in teaching math!

Maria Miller, the author

Games and Activities

Subtraction Challenge

You need: A standard deck of playing cards from which you remove the face cards.

Game Play: In each round, each player is dealt two cards face up, and has to calculate their difference (subtract). The player with the highest difference (highest answer) gets all the cards played in that round, including from the other players, and puts them in his personal pile. If two or more players have the same answer, then those players get an additional two cards, and use those to resolve the tie.

After enough rounds have been played to use all of the cards, the player with the most cards wins.

Number Line Activity

You need: A six-sided dice and two play markers. Also, draw a number line from 0 to 15 on a blank paper.

Mark number 10 in a special way. For example, you can make it bolder or bigger, or circle it.

Game Play: Place both markers at zero on the number line. The goal is to reach number 10. On your turn, roll the dice, and say an addition sentence starting with the position where your marker is, adding to it what the dice shows. For example, if your marker is at 2, and you roll a 5, you say "$2 + 5 = 7$", and then move your marker to seven. Once your marker is past 10, you say a subtraction sentence instead. For example, if your marker is at 12, and you roll a 4, you will say, "$12 - 4 = 8$", and move your marker to 4. The winner is the person who gets their game piece to land at 10.

It is important to say the addition/subtraction sentence before moving the piece. Otherwise, this activity amounts to counting steps on a number line and is reduced to a counting activity.

Note: If a child cannot do a particular subtraction or addition, they can use the number line to figure out the answer, before moving their piece.

Let's Make a Difference!

You need: A large deck of number cards with numbers 1-10.

Preparation: Deal six cards to each player. Place the rest face down in a pile in the middle of the table.

Game play: At your turn, check if you can make a subtraction sentence using any three cards in your hand. If so, say that subtraction aloud, and put those three cards to your personal pile. If not, take *one* card from the middle pile, and check again. If you still cannot make a subtraction, your turn ends. At the end of your turn, draw enough cards from the middle pile so that you once again have at least six cards in your hand (if you already have six or more, don't draw more at this point). The turn then passes to the next player.

Play in this manner until there are no more cards in the middle pile. At that point, the player with most cards in their personal pile wins.

Note: Essentially, players can check if they have three cards that make an *addition* sentence. However, it is best if the child notices that on his/her own.

Variation: For an easier game, remove 10s, 9s, and perhaps even 8s from the deck.

Games and Activities at Math Mammoth Practice Zone

Connection between Addition and Subtraction
https://www.mathmammoth.com/practice/add-sub-connection

Fact Families
https://www.mathmammoth.com/practice/fact-families

Subtraction Hidden Picture Game
https://www.mathmammoth.com/practice/mystery-picture-subtraction

Number Bonds
https://www.mathmammoth.com/practice/number-bonds

Fruity Math
https://www.mathmammoth.com/practice/fruity-math

Further Resources on the Internet

We have compiled a list of external Internet resources that match the topics in this book. This list of links includes web pages that offer:

- **online practice** for concepts;

- online **games**, or occasionally, printable games;

- **animations** and interactive **illustrations** of math concepts;

- **articles** that teach a math concept.

We heartily recommend you take a look at the list. Many of our customers love using these resources to supplement the bookwork. You can use the resources as you see fit for extra practice, to illustrate a concept better, and even just for some fun. Enjoy!

https://l.mathmammoth.com/blue/subtraction1

Scan me

Subtraction is "Taking Away"

From five bananas we take away three.   There are two left.

$$5 - 3 \qquad = \qquad 2$$

Five minus three   equals   two

From six carrots we take away one.   There are five left.

$$6 - 1 \qquad = \qquad 5$$

Six minus one   equals   five

1. Cross out objects. How many are left? Read each subtraction sentence aloud using the words "minus" and "equals".

10 in total. Cross out 4. How many are left?	6 in total. Cross out 6. How many are left?	8 in total. Cross out 2. How many are left?
a. $10 - 4 =$ _____	**b.** $6 - 6 =$ _____	**c.** $8 - 2 =$ _____
8 in total, cross out 7.	8 in total, cross out 6.	7 in total, cross out 3.
d. $8 - 7 =$ _____	**e.** $8 - 6 =$ _____	**f.** $7 - 3 =$ _____

9 in total, cross out 3.	6 in total, cross out 2.	10 in total, cross out 7.
g. $9 - 3 = $ _____	h. $6 - 2 = $ _____	i. $10 - 7 = $ _____
Cross out 5.	Cross out 4.	Cross out 5.
j. $9 - 5 = $ _____	k. $8 - 4 = $ _____	l. $10 - 5 = $ _____
m. $6 - 3 = $ _____	n. $6 - 5 = $ _____	o. $5 - 5 = $ _____

2. Subtract. Cover the crossed-out objects with your finger to see how many are left.
   Read each sentence using the words "minus" and "equals".

a. $5 - 1 = $ _____	b. $6 - 3 = $ _____	c. $5 - 3 = $ _____
d. $4 - 1 = $ _____	e. $5 - 2 = $ _____	f. $6 - 2 = $ _____
g. $9 - 3 = $ _____	h. $7 - 1 = $ _____	i. $7 - 2 = $ _____
j. $8 - 2 = $ _____	k. $7 - 3 = $ _____	l. $8 - 4 = $ _____

3. Draw small circles to illustrate the numbers and cross out some of them to match the subtraction problem.

a. $8 - 3 =$ _____	**b.** $5 - 1 =$ _____	**c.** $10 - 7 =$ _____
d. $7 - 2 =$ _____	**e.** $10 - 1 =$ _____	**f.** $9 - 7 =$ _____
g. $6 - 3 =$ _____	**h.** $7 - 3 =$ _____	**i.** $10 - 4 =$ _____
j. $9 - 5 =$ _____	**k.** $10 - 6 =$ _____	**l.** $6 - 4 =$ _____

4. Write a subtraction sentence to match the picture.

a. _____ – _____ = _____	**b.** _____ – _____ = _____
c. _____ – _____ = _____	**d.** _____ – _____ = _____
e. _____ – _____ = _____	**f.** _____ – _____ = _____

Count Down to Subtract

Count 4 steps backwards from 7. You end up at 3.

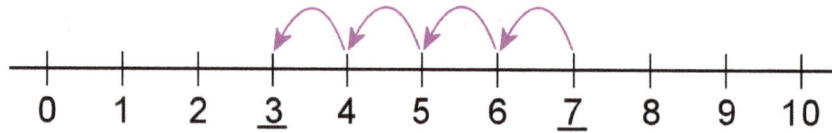

$$7 - 4 = 3$$

"Seven"   "Six, five, four, three."

Start.   Count down FOUR steps.

You can also use a single arrow:

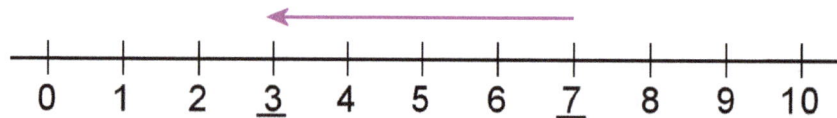

1. Draw steps (or a single arrow) to illustrate the subtraction sentence.

a. $7 - 2 =$ _____

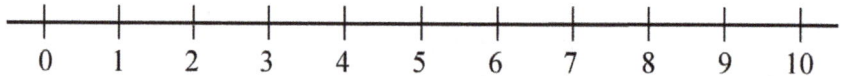

b. $8 - 4 =$ _____

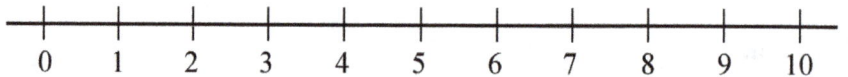

c. $6 - 5 =$ _____

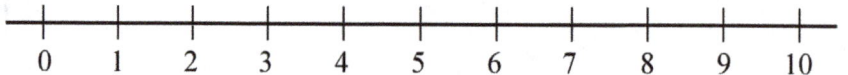

d. $9 - 3 =$ _____

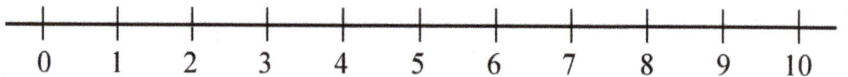

e. $10 - 3 =$ _____

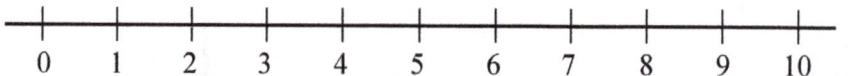

2. Write the subtraction sentence that the arrows illustrate.

a. _____ − _____ = 2

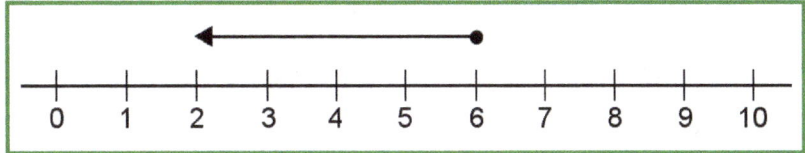

b. _____ − _____ = 3

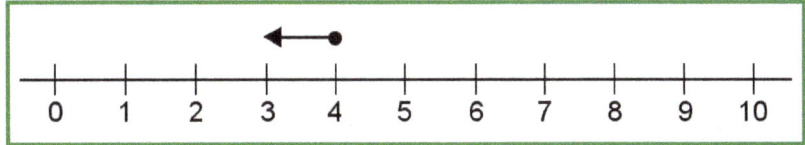

c. _____ − _____ = 2

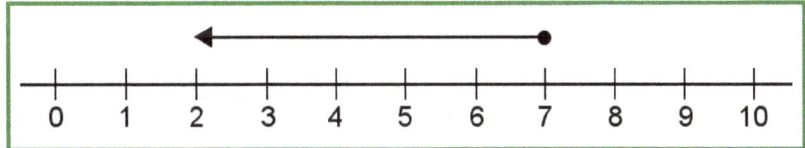

d. _____ − _____ = 5

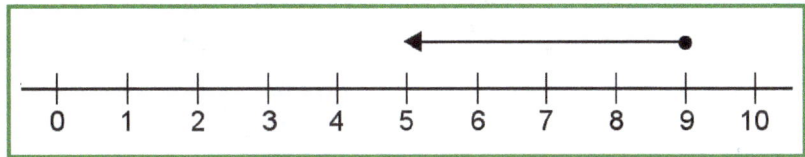

3. Draw an arrow for the subtraction sentence and solve.

a. $10 - 5 =$ _____

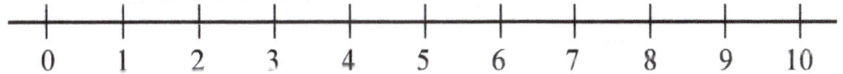

b. $7 - 6 =$ _____

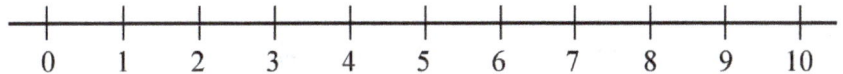

c. $4 - 4 =$ _____

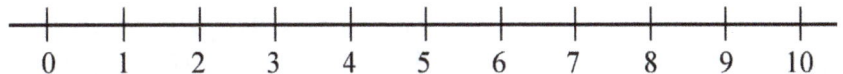

d. $8 - 4 =$ _____

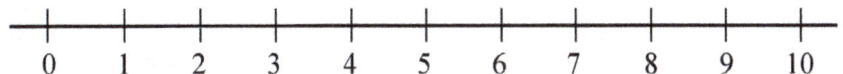

e. $10 - 1 =$ _____

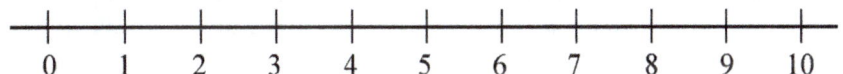

f. $7 - 5 =$ _____

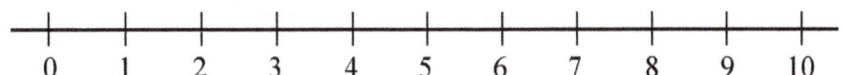

4. Write the previous and the next numbers.

a. _____ , 5 , _____   b. _____ , 2 , _____   c. _____ , 8 , _____

d. _____ , 6 , _____   e. _____ , 4 , _____   f. _____ , 9 , _____

5. Write the two previous numbers.

a. _____ , _____ , 7   b. _____ , _____ , 4   c. _____ , _____ , 10

d. _____ , _____ , 6   e. _____ , _____ , 2   f. _____ , _____ , 8

To subtract without any pictures or objects, you can count down in your head.
Count down as many steps as what you need to take away or subtract.

8	− 3	= 5		10	− 2	= 8
"Eight"	"Seven, six, five."			"Ten"	"Nine, eight."	
Start.	Count down THREE steps.			Start.	Count down TWO steps.	

6. Subtract one. The answer is just the previous number!

a.	b.	c.	d.
$6 - 1 =$ _____	$5 - 1 =$ _____	$9 - 1 =$ _____	$4 - 1 =$ _____
$8 - 1 =$ _____	$7 - 1 =$ _____	$3 - 1 =$ _____	$10 - 1 =$ _____

7. Subtract 2 or 3. You can count down. Compare the problems.

a.	b.	c.	d.
$6 - 2 =$ _____	$9 - 2 =$ _____	$7 - 2 =$ _____	$10 - 2 =$ _____
$6 - 3 =$ _____	$9 - 3 =$ _____	$7 - 3 =$ _____	$10 - 3 =$ _____

8. Solve the word problems. Write a subtraction sentence for each.

a. There were 7 birds in a tree. Three flew away. How many are left?	b. Mom has 10 silver plates in the cupboard. She took out four. How many are still in the cupboard?
c. All 9 girls in the class were jumping rope. Then four of them left. How many kept on jumping?	d. Josh took five of his 10 toy cars to a friend's house. How many cars did he leave at home?
e. Of her eight puzzles, Fanny put three in the closet. How many were left to play with?	f. Tina had 6 bunches of flowers. She sold six of them. How many were left?

9. Do these problems if you know the numbers past 10. The number line will help.

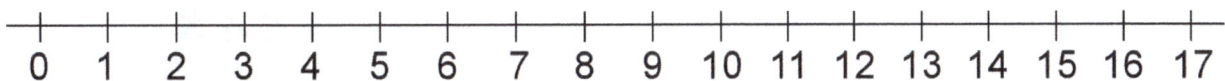

0 1 2 3 4 5 6 7 8 9 10 11 12 13 14 15 16 17

a. $14 - 2 =$ _____ $14 - 4 =$ _____	b. $16 - 1 =$ _____ $16 - 3 =$ _____	c. $11 - 3 =$ _____ $11 - 4 =$ _____
d. $17 - 2 =$ _____ $17 - 3 =$ _____	e. $12 - 1 =$ _____ $12 - 2 =$ _____	f. $13 - 2 =$ _____ $13 - 3 =$ _____

Subtraction and Addition in the Same Picture

How many colored circles?
How many white ones?

⬤⬤⬤⬤◯◯◯◯◯◯

__4__ + __6__ = 10

⬤⬤⬤◯◯◯◯

__3__ + __4__ = 7

Cover the colored circles.
Write a subtraction sentence.

⬤⬤⬤⬤◯◯◯◯◯◯

10 – __4__ = __6__

Cover the colored circles.

⬤⬤⬤◯◯◯◯

7 – __3__ = __4__

1. Make an addition sentence and a subtraction sentence from the same picture.

a. ⬤⬤⬤◯◯◯◯

_____ + _____ = _____

7 – _____ = _____

b. �✦✦✦✦✦✦✦

_____ + _____ = _____

6 – _____ = _____

c. ⬤◯◯◯◯

_____ + _____ = _____

5 – _____ = _____

d. ✦✦✦✦✦✦

_____ + _____ = _____

6 – _____ = _____

e. ▢▢▢▢▢🟥🟥🟥

_____ + _____ = _____

8 – _____ = _____

f. ▢▢▢🟥🟥🟥

_____ + _____ = _____

6 – _____ = _____

2. Make an addition sentence and a subtraction sentence for the same picture.

a.

_____ + _____ = _____

_____ − _____ = _____

b.

_____ + _____ = _____

_____ − _____ = _____

c.

_____ + _____ = _____

_____ − _____ = _____

d.

_____ + _____ = _____

_____ − _____ = _____

3. In each problem, draw circles and then color them to fit the addition sentence.
   Then cover the **COLORED** circles and make a subtraction sentence.

a. 7 + 1 = _____

_____ − _____ = _____

b. 6 + 3 = _____

_____ − _____ = _____

c. 2 + 3 = _____

_____ − _____ = _____

d. 2 + 5 = _____

_____ − _____ = _____

e. 7 + 4 = _____

_____ − _____ = _____

f. 3 + 3 = _____

_____ − _____ = _____

4. Cover the colored objects, and write a subtraction sentence to fit the picture.

a.

_____ − _____ = _____

b.

_____ − _____ = _____

c.

_____ − _____ = _____

d.

_____ − _____ = _____

e.

_____ − _____ = _____

f.

_____ − _____ = _____

g.

_____ − _____ = _____

h.

_____ − _____ = _____

5. In each problem, draw some circles and color them to fit the addition sentence.
   Then cover the **COLORED** circles and make a subtraction sentence.

a. $9 + 1 = $ _____

_____ − _____ = _____

b. $7 + 2 = $ _____

_____ − _____ = _____

c. $10 + 4 = $ ___

_____ − _____ = _____

d. $10 + 2 = $ ___

_____ − _____ = _____

6. Draw circles to fit the subtraction sentence. Write an addition sentence, too.

a. $9 - 4 =$ ___

___ $+$ ___ $=$ ___

b. $10 - 5 =$ ___

___ $+$ ___ $=$ ___

c. $8 - 5 =$ ___

___ $+$ ___ $=$ ___

d. $8 - 4 =$ ___

___ $+$ ___ $=$ ___

e. $7 - 4 =$ ___

___ $+$ ___ $=$ ___

f. $9 - 8 =$ ___

___ $+$ ___ $=$ ___

Puzzle Corner Subtract or add. Do not write the answers; just solve them in your head. Then compare, and write $<$, $>$ or $=$.

3 ☐ $3 - 1$ $6 + 5$ ☐ 6 10 ☐ $10 - 1$

$9 - 7$ ☐ $8 - 7$ $6 - 4$ ☐ $2 + 3$ $8 - 5$ ☐ $5 + 3$

$5 + 2$ ☐ $8 + 2$ $10 - 1$ ☐ $10 - 3$ $7 - 4$ ☐ $8 - 5$

$10 - 2$ ☐ $8 - 2$ $10 + 0$ ☐ $10 - 0$ $8 - 1$ ☐ $8 + 1$

When Can You Subtract?

Sometimes we cannot really do the subtraction problem. For example:

3 – 0 = _____   3 – 4 = ??

3 – 1 = _____   3 – 5 = ??

△ △ △

3 – 2 = _____   3 – 6 = ??

3 – 3 = _____   We cannot take away
four or five or six objects
These we can do!   when there are only three!

1. Write the subtraction problems you *can* do when there are...

...four triangles	...six circles	...five sticks
△ △ △ △	○ ○ ○ ○ ○ ○	❘❘❘❘❘
4 – 0 = ___	6 – 0 = ___	___ – ___ = ___
4 – 1 = ___	___ – ___ = ___	___ – ___ = ___
4 – 2 = ___	___ – ___ = ___	___ – ___ = ___
___ – ___ = ___	___ – ___ = ___	___ – ___ = ___
___ – ___ = ___	___ – ___ = ___	___ – ___ = ___
	___ – ___ = ___	___ – ___ = ___
	___ – ___ = ___	

You cannot do a subtraction problem in whole numbers (0, 1, 2, 3, ...)
when the second number is _____ than the first number.

2. Count down to subtract.

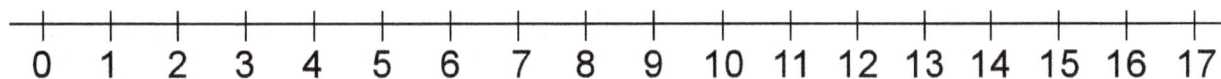

$$0 \quad 1 \quad 2 \quad 3 \quad 4 \quad 5 \quad 6 \quad 7 \quad 8 \quad 9 \quad 10 \quad 11 \quad 12 \quad 13 \quad 14 \quad 15 \quad 16 \quad 17$$

a.	b.	c.	d.
$7 - 1 =$ _____	$9 - 1 =$ _____	$10 - 1 =$ _____	$12 - 1 =$ _____
$7 - 2 =$ _____	$9 - 2 =$ _____	$10 - 2 =$ _____	$12 - 2 =$ _____
$7 - 3 =$ _____	$9 - 3 =$ _____	$10 - 3 =$ _____	$12 - 3 =$ _____

3. Continue the patterns as long as you can!

a.	b.	c.
$7 - 0 =$ ___	$10 - 5 =$ ___	$8 - 2 =$ ___
$7 - 1 =$ ___	$9 - 5 =$ ___	$7 - 2 =$ ___
$7 - 2 =$ ___	$8 - 5 =$ ___	$6 - 2 =$ ___
$7 -$ ___ $=$ ___	___ $- 5 =$ ___	___ $-$ ___ $=$ ___
___ $-$ ___ $=$ ___	___ $-$ ___ $=$ ___	___ $-$ ___ $=$ ___
___ $-$ ___ $=$ ___	___ $-$ ___ $=$ ___	___ $-$ ___ $=$ ___
___ $-$ ___ $=$ ___	___ $-$ ___ $=$ ___	___ $-$ ___ $=$ ___
___ $-$ ___ $=$ ___	___ $-$ ___ $=$ ___	___ $-$ ___ $=$ ___
___ $-$ ___ $=$ ___	___ $-$ ___ $=$ ___	___ $-$ ___ $=$ ___

4. Find the subtractions where you can't take away that many, and cross them out.
   You *don't* have to write the answers!

$4-0$	$7-7$	$5-6$	$3-6$	$4-4$	$3-10$
$4-5$	$7-9$	$10-1$	$3-4$	$2-4$	$4-3$

5. Are these subtractions right? Circle true or false.

a. $7-1=8$ true *or* false	**d.** $5-2=6$ true *or* false
b. $9-2=7$ true *or* false	**e.** $10-8=1$ true *or* false
c. $10-5=4$ true *or* false	**f.** $6-3=3$ true *or* false

6. For each problem, answer these questions: Can the child buy the item? Yes or no.
   If yes, how much money will she/he have left?
   If not, how much more money would she/he need to buy the item?

a. Jennie has three dollars. She wants to buy a doll that costs five dollars.	**b.** Jessie has $5. He wants a ball that costs $2.
c. Lola has seven dollars. She wants to buy a Lego set that costs four dollars.	**d.** Marvin has $5. He wants a book that costs $6.
e. Jack has eight dollars. He wants to buy a construction set that costs ten dollars.	**f.** Mary has seven dollars. She wants a car that costs two dollars.
g. Faye has $12. She wants to buy a game that costs $4.	**h.** Anthony has ten dollars. He wants a game that costs fifteen dollars.

7. Subtraction can be written this way too! Write the answer below the line.

a.
$$\begin{array}{r} 10 \\ - 3 \\ \hline \end{array}$$

b.
$$\begin{array}{r} 8 \\ - 7 \\ \hline \end{array}$$

c.
$$\begin{array}{r} 6 \\ - 5 \\ \hline \end{array}$$

d.
$$\begin{array}{r} 8 \\ - 6 \\ \hline \end{array}$$

e.
$$\begin{array}{r} 8 \\ - 0 \\ \hline \end{array}$$

f.
$$\begin{array}{r} 7 \\ - 7 \\ \hline \end{array}$$

g.
$$\begin{array}{r} 7 \\ - 6 \\ \hline \end{array}$$

h.
$$\begin{array}{r} 6 \\ - 6 \\ \hline \end{array}$$

i.
$$\begin{array}{r} 6 \\ - 1 \\ \hline \end{array}$$

j.
$$\begin{array}{r} 9 \\ - 4 \\ \hline \end{array}$$

k.
$$\begin{array}{r} 10 \\ - 8 \\ \hline \end{array}$$

l.
$$\begin{array}{r} 4 \\ - 0 \\ \hline \end{array}$$

m.
$$\begin{array}{r} 6 \\ - 4 \\ \hline \end{array}$$

n.
$$\begin{array}{r} 7 \\ - 2 \\ \hline \end{array}$$

o.
$$\begin{array}{r} 9 \\ - 3 \\ \hline \end{array}$$

8. If the answer is four, color the box red. If the answer is five, color the box orange. If the answer is ten, color the box yellow. And, if you can't subtract, color the box light blue.

10 − 6	6 − 7	5 − 6	5 − 10	7 − 3
1 − 6	10 − 5	9 − 10	8 − 3	1 − 3
7 − 9	1 − 2	10 − 0	3 − 5	6 − 9
4 − 8	6 − 1	0 − 9	9 − 4	7 − 8
8 − 4	0 − 4	2 − 6	6 − 9	9 − 5

Two Subtractions from One Addition

From one addition fact, you can make TWO subtraction facts:

3 + 4 = 7

⇩

7 – 4 = 3   or      7 – 3 = 4

Take away the carrots in the *bottom* row.

Take away the carrots in the *top* row.

Sometimes the two subtraction facts are the same:

2 + 2 = 4

⇩

4 – 2 = 2   or      4 – 2 = 2

Take away the carrots in the *bottom* row.

Take away the carrots in the *top* row.

1. Write one addition and two subtraction sentences. First subtract the things in the bottom row then the ones in the top row.

a. 1 + 3 = ___4___

   4 – 3 = _____

or 4 – 1 = _____

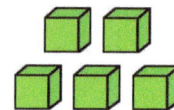

b. 2 + 3 = ___5___

   5 – _____ = _____

or 5 – _____ = _____

c. _____ + _____ = _____

   _____ – _____ = _____

or _____ – _____ = _____

d. _____ + _____ = _____

   _____ – _____ = _____

or _____ – _____ = _____

e. _____ + _____ = _____

_____ − _____ = _____

or _____ − _____ = _____

f. _____ + _____ = _____

_____ − _____ = _____

or _____ − _____ = _____

g. _____ + _____ = _____

_____ − _____ = _____

or _____ − _____ = _____

h. _____ + _____ = _____

_____ − _____ = _____

or _____ − _____ = _____

i. _____ + _____ = _____

_____ − _____ = _____

or _____ − _____ = _____

j. _____ + _____ = _____

_____ − _____ = _____

or _____ − _____ = _____

k. _____ + _____ = _____

_____ − _____ = _____

or _____ − _____ = _____

l. _____ + _____ = _____

_____ − _____ = _____

or _____ − _____ = _____

2. Complete the addition fact and the subtraction facts. Draw Xs in two groups.

XXXXXX **X**	**XXXX** **XXX**
a. $6 + 1 = 7$ $7 - 1 = 6$ $7 - 6 = 1$	**b.** $4 + 3 = 7$ $7 - \underline{\quad} = \underline{\quad}$ $7 - \underline{\quad} = \underline{\quad}$
c. $7 + 1 = \underline{\quad}$ $\underline{\quad} - \underline{\quad} = \underline{\quad}$ $\underline{\quad} - \underline{\quad} = \underline{\quad}$	**d.** $5 + 4 = \underline{\quad}$ $\underline{\quad} - \underline{\quad} = \underline{\quad}$ $\underline{\quad} - \underline{\quad} = \underline{\quad}$
e. $2 + 6 = \underline{\quad}$ $\underline{\quad} - \underline{\quad} = \underline{\quad}$ $\underline{\quad} - \underline{\quad} = \underline{\quad}$	**f.** $3 + 5 = \underline{\quad}$ $\underline{\quad} - \underline{\quad} = \underline{\quad}$ $\underline{\quad} - \underline{\quad} = \underline{\quad}$
g. $1 + 8 = \underline{\quad}$ $\underline{\quad} - \underline{\quad} = \underline{\quad}$ $\underline{\quad} - \underline{\quad} = \underline{\quad}$	**h.** $5 + 5 = \underline{\quad}$ $\underline{\quad} - \underline{\quad} = \underline{\quad}$ $\underline{\quad} - \underline{\quad} = \underline{\quad}$

Two Parts — One Total

There are ten marbles. Some of them are blue and seven are yellow. How many are blue?

You can write an addition sentence. You can ALSO write a subtraction sentence, even though nothing is taken away.

$\underline{\qquad} + 7 = 10$

$10 - 7 = \underline{\qquad}$

Cover part of the total (the yellow marbles), and you will see the other part (the blue marbles).

There are five blue marbles and some white marbles in a bag. There is a total of nine marbles. How many are white?

Draw the marbles. Write an addition sentence AND a subtraction sentence.

$\underline{\qquad} + \underline{\qquad} = \underline{\qquad}$

$\underline{\qquad} - \underline{\qquad} = \underline{\qquad}$

1. Solve the word problems. Write an addition sentence AND a subtraction sentence.

a. Mother put some blue and some red flowers in a vase. Jen counted five red ones and a total of ten. How many of the flowers are blue?

$\underline{\qquad} + \underline{\qquad} = \underline{\qquad}$

$\underline{\qquad} - \underline{\qquad} = \underline{\qquad}$

b. There are nine children on a team, and four of them are boys. How many are girls?

$\underline{\qquad} + \underline{\qquad} = \underline{\qquad}$

$\underline{\qquad} - \underline{\qquad} = \underline{\qquad}$

c. Jack has ten socks in his basket. Eight of them are white, and the rest are black. How many are black?

_____ + _____ = _____

_____ – _____ = _____

d. Mary saw eight chairs on the lawn, and two had blown over. How many were still standing upright?

_____ + _____ = _____

_____ – _____ = _____

2. For each picture, make a word problem that is solved by subtraction.

a.

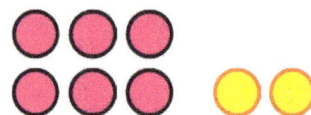

b.

3. Write an addition sentence for the pictures.

a. _____ + _____ + _____ = _____

b. _____ + _____ + _____ = _____

c. _____ + _____ + _____ = _____

d. _____ + _____ + _____ = _____

4. Draw the missing marbles to match the addition sentence.

a. 3 + 2 + _____ = 8

b. 1 + 5 + _____ = 10

5. Draw a picture to solve these problems.

a. Jane had some red, some blue, and some yellow roses in a vase. Two of the roses were blue, and two were red. If she had a total of ten roses, how many of them were yellow?

b. Seven birds sat in a tree. One of them was black, two were blue, and the rest were brown. How many were brown?

c. Mary has two long pencils and two medium-sized ones. The rest of her pencils are short If she owns nine pencils in all, how many of her pencils are short?

Fact Families

Two addition facts and two subtraction facts form a <u>fact family</u> if they use the same three numbers.

For example, from 5, 3, and 2 we get the fact family on the right:

5

●● / ●●●

$2 + 3 = 5$ $5 - 3 = 2$

$3 + 2 = 5$ $5 - 2 = 3$

1. Write the fact families that match the pictures.

a. 6

● / ●●●
　　 ●●

$1 + 5 = 6$

$5 + 1 = 6$

$6 - \underline{\hspace{1.5em}} = \underline{\hspace{1.5em}}$

$6 - \underline{\hspace{1.5em}} = \underline{\hspace{1.5em}}$

b. 8

●● / ●●●
●　　 ●●

$\underline{\hspace{1.5em}} + \underline{\hspace{1.5em}} = \underline{\hspace{1.5em}}$

$\underline{\hspace{1.5em}} + \underline{\hspace{1.5em}} = \underline{\hspace{1.5em}}$

$\underline{\hspace{1.5em}} - \underline{\hspace{1.5em}} = \underline{\hspace{1.5em}}$

$\underline{\hspace{1.5em}} - \underline{\hspace{1.5em}} = \underline{\hspace{1.5em}}$

c. 9

●● / ●●●
●　　 ●●●

$\underline{\hspace{1.5em}} + \underline{\hspace{1.5em}} = \underline{\hspace{1.5em}}$

$\underline{\hspace{1.5em}} + \underline{\hspace{1.5em}} = \underline{\hspace{1.5em}}$

$\underline{\hspace{1.5em}} - \underline{\hspace{1.5em}} = \underline{\hspace{1.5em}}$

$\underline{\hspace{1.5em}} - \underline{\hspace{1.5em}} = \underline{\hspace{1.5em}}$

d. 10

●●● / ●
●●●●● 　 ●●

$\underline{\hspace{1.5em}} + \underline{\hspace{1.5em}} = \underline{\hspace{1.5em}}$

$\underline{\hspace{1.5em}} + \underline{\hspace{1.5em}} = \underline{\hspace{1.5em}}$

$\underline{\hspace{1.5em}} - \underline{\hspace{1.5em}} = \underline{\hspace{1.5em}}$

$\underline{\hspace{1.5em}} - \underline{\hspace{1.5em}} = \underline{\hspace{1.5em}}$

2. Draw circles and write four *different* fact families for which the sum is 7.

a. 7
/

_____ + _____ = _____

_____ + _____ = _____

_____ − _____ = _____

_____ − _____ = _____

b. 7
/

_____ + _____ = _____

_____ + _____ = _____

_____ − _____ = _____

_____ − _____ = _____

c. 7
/

_____ + _____ = _____

_____ + _____ = _____

_____ − _____ = _____

_____ − _____ = _____

d. 7
/

_____ + _____ = _____

_____ + _____ = _____

_____ − _____ = _____

_____ − _____ = _____

3. Ann and Joe solved some math problems that had missing (unknown) numbers. Play math teacher. Check their work and correct any mistakes that they made.

Ann:

a. $8 - 4 = \boxed{4}$

b. $\boxed{6} - 4 = 1$

c. $5 - \boxed{2} = 2$

Joe:

d. $\boxed{3} - 4 = 1$

e. $10 - 8 = \boxed{1}$

f. $8 - \boxed{7} = 1$

4. Make the four facts for the given numbers.

a. Numbers: 5, 3, 2

$$2 + 3 = 5$$

$$3 + 2 = 5$$

$$5 - 3 = 2$$

$$5 - 2 = 3$$

b. Numbers: 9, 4, 5

$$4 + 5 = 9$$

$$\underline{\quad} + \underline{\quad} = 9$$

$$9 - 4 = \underline{\quad}$$

$$9 - \underline{\quad} = \underline{\quad}$$

c. Numbers: 4, 0, 4

$$4 + 0 = 4$$

$$\underline{\quad} + \underline{\quad} = \underline{\quad}$$

$$4 - 0 = \underline{\quad}$$

$$\underline{\quad} - \underline{\quad} = \underline{\quad}$$

d. Numbers: 10, 3, 7

$$\underline{\quad} + \underline{\quad} = \underline{\quad}$$

$$\underline{\quad} + \underline{\quad} = \underline{\quad}$$

$$\underline{\quad} - \underline{\quad} = \underline{\quad}$$

$$\underline{\quad} - \underline{\quad} = \underline{\quad}$$

e. Numbers: 10, \underline{\quad}, 8

$$\underline{\quad} + \underline{\quad} = \underline{\quad}$$

$$\underline{\quad} + \underline{\quad} = \underline{\quad}$$

$$\underline{\quad} - \underline{\quad} = \underline{\quad}$$

$$\underline{\quad} - \underline{\quad} = \underline{\quad}$$

f. Numbers: 6, 0, \underline{\quad}

$$\underline{\quad} + \underline{\quad} = \underline{\quad}$$

$$\underline{\quad} + \underline{\quad} = \underline{\quad}$$

$$\underline{\quad} - \underline{\quad} = \underline{\quad}$$

$$\underline{\quad} - \underline{\quad} = \underline{\quad}$$

Sometimes the two addition facts are the same. When that happens, the two subtraction facts will also be the same.

For example, with 8, 4, and 4, we only get one addition fact and one subtraction fact.

$$4 + 4 = 8 \qquad 8 - 4 = 4$$

$$(4 + 4 = 8) \qquad (8 - 4 = 4)$$

5. Write the fact families.

a. Numbers: 10, 5, 5

_____ + _____ = _____

_____ + _____ = _____

_____ − _____ = _____

_____ − _____ = _____

b. Numbers: 9, 1, 8

_____ + _____ = _____

_____ + _____ = _____

_____ − _____ = _____

_____ − _____ = _____

c. Numbers: 6, 3, _____

_____ + _____ = _____

_____ + _____ = _____

_____ − _____ = _____

_____ − _____ = _____

d. Numbers: 7, 1, _____

_____ + _____ = _____

_____ + _____ = _____

_____ − _____ = _____

_____ − _____ = _____

Puzzle Corner For each equation, write the number that the ◯, ☐, or △ represents.

$$\bigcirc - 4 = 5 \qquad 5 - \square = 4 \qquad \triangle + 4 = 5$$

$$\bigcirc - 2 = 3 \qquad 7 - \square = 1 \qquad \triangle + 2 = 3$$

$$\bigcirc - 7 = 1 \qquad 10 - \square = 5 \qquad \triangle + 7 = 10$$

33

How Many More?

⚫⚫⚫ Henry | | | ⚫⚫⚫⚫⚫ Cindy Cindy has more marbles. *How many more?* Match three marbles from each box. Two marbles did not get matched, so Cindy has <u>2 more marbles than</u> Henry. (Henry has <u>2 fewer marbles than</u> Cindy.)	⚫⚫⚫⚫⚫ Henry | | ⚫⚫ Cindy Cindy has fewer marbles. *How many fewer?* Match two marbles from each box. Four marbles did not get matched, so Cindy has <u>4 fewer marbles than</u> Henry. (Henry has <u>4 more marbles than</u> Cindy.)

1. Fill in how many more or how many fewer marbles one child has than another.

⚫⚫⚫⚫⚫⚫ Jane

⚫⚫⚫⚫ Jim

a. Jane has _____ more than Jim.

Jim has _____ fewer than Jane.

⚫⚫⚫⚫⚫⚫⚫ Mark

⚫⚫⚫⚫ Mary

b. Mark has _____ more than Mary.

Mary has _____ fewer than Mark.

⚫⚫⚫⚫ Ann

⚫⚫⚫⚫⚫⚫ Liz

c. Liz has _____ more than Ann.

Ann has _____ fewer than Liz.

⚫ Faye

⚫⚫⚫⚫⚫⚫ Sam

d. Sam has _____ more than Faye.

Faye has _____ fewer than Sam.

⚫⚫ Peter

⚫⚫⚫⚫⚫⚫⚫⚫ Frank

e. Frank has _____ more than Peter.

Peter has _____ fewer than Frank.

⚫⚫⚫⚫⚫⚫ Susan

⚫ Bill

f. Susan has _____ more than Bill.

Bill has _____ fewer than Susan.

2. Now it's your turn to draw. Draw marbles for the child that has none.

[ ]	Jane
🔵🔵🔵🔵🔵	Jim

a. Jane has 3 more than Jim.

🔵🔵	Mark
[ ]	Mary

b. Mary has 4 more than Mark.

[ ]	Eric
🔵🔵🔵	Bill

c. Eric has 2 fewer than Bill.

🔵🔵🔵🔵🔵🔵🔵	Jack
[ ]	Jane

d. Jane has 5 fewer than Jack.

[ ]	Bill
🔵🔵🔵🔵🔵🔵	Greg

e. Greg has 1 more than Bill.

🟢🟢🟢🟢🟢🟢🟢🟢🟢	Lucy
[ ]	Liz

f. Lucy has 5 more than Liz.

[ ]	Ed
🟡🟡🟡🟡🟡	Sally

g. Sally has 2 fewer than Ed.

🟡🟡🟡	Ann
[ ]	Mary

h. Ann has 4 fewer than Mary.

[ ]	Sue
🔵🔵🔵🔵🔵🔵🔵	Ben

i. Ben has five more than Sue.

🟢🟢🟢🟢	Jill
[ ]	Mary

j. Jill has five fewer than Mary.

3. It is still your turn to draw. You can decide how many marbles the children have.

[ ] Jane	[ ] Mark
[ ] Jim	[ ] Mary
a. Jane has 5 more than Jim.	**b.** Mary has 2 more than Mark.
[ ] Eric	[ ] Jack
[ ] Bill	[ ] Jane
c. Eric has 6 fewer than Bill.	**d.** Jane has 7 more than Jack.
[ ] Bill	[ ] Lucy
[ ] Greg	[ ] Liz
e. Greg has 2 fewer than Bill.	**f.** Lucy has 8 more than Liz.

4. Solve these problems.

a. Ed has five cards, and Jack has seven. How many more cards does Jack have than Ed?

b. John is 8 years old and Jack is 5. How many years older is John?

c. Annie is 10 years old and Beth is 8. How many years younger is Beth?

d. Ruth had 9 dolls, and Tina had 4 dolls. How many fewer did Tina have?

"How Many More" Problems and Differences

How <u>many more</u> emails does Jane have than John?

Draw more emails for John so that the children have the same amount. I drew _____ more emails.

You can write an addition for a "*how many more*" problem.

John

Jane

$2 + \underline{\hspace{1cm}} = 5$

Read: "2 and *how many more* makes five?"

1. Draw more. Read the addition sentences.

a. $1 + \underline{\hspace{1cm}} = 4$ "1 and how many more makes 4?"	**b.** $3 + \underline{\hspace{1cm}} = 6$ "3 and how many more makes 6?"	**c.** $3 + \underline{\hspace{1cm}} = 7$ "3 and how many more makes 7?"
d. $4 + \underline{\hspace{1cm}} = 10$	**e.** $2 + \underline{\hspace{1cm}} = 8$	**f.** $1 + \underline{\hspace{1cm}} = 5$

The problem $\underline{\hspace{1cm}} + 2 = 5$ is also read "2 and *how many more* makes five?"

2. Draw more. Solve.

a. $\underline{\hspace{1cm}} + 5 = 7$ 5 and how many more makes 7?	**b.** $\underline{\hspace{1cm}} + 2 = 6$ 2 and how many more makes 6?	**c.** $\underline{\hspace{1cm}} + 1 = 8$ 1 and how many more makes 8?
d. $\underline{\hspace{1cm}} + 2 = 7$	**e.** $\underline{\hspace{1cm}} + 4 = 7$	**f.** $\underline{\hspace{1cm}} + 3 = 6$

How many more problems are DIFFERENCE problems. The difference of two numbers means <u>how far</u> the two numbers are from each other.

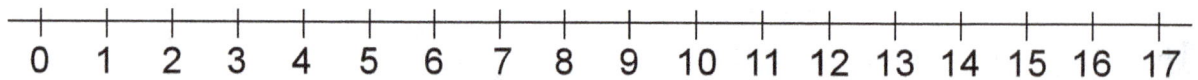

```
+---+---+---+---+---+---+---+---+---+---+---+---+---+---+---+---+---+
0   1   2   3   4   5   6   7   8   9  10  11  12  13  14  15  16  17
```

How far is 3 from 7?

How much difference is there between 3 and 7?

3 and how many more make 7?

$3 + \rule{1cm}{0.15mm} = 7$

The answer to all these is 4.

How far is 8 from 13?

How much difference is there between 8 and 13?

8 and how many more make 13?

$8 + \rule{1cm}{0.15mm} = 13$

The answer to all these is 5.

3. What is the difference between the numbers? Take steps on the number line.

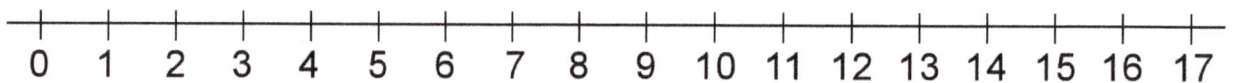

```
+---+---+---+---+---+---+---+---+---+---+---+---+---+---+---+---+---+
0   1   2   3   4   5   6   7   8   9  10  11  12  13  14  15  16  17
```

a. from 6 to 10	**b.** from 5 to 8	**c.** from 7 to 11	**d.** from 5 to 5	**e.** from 1 to 10
_____ steps	_____ steps	_____ steps	_____ steps	_____ steps

4. Find the difference between the numbers. "Travel" on the number line!

From	8	4	1	3	6	10	8	9
To	10	10	9	1	5	5	12	15
Difference								

5. Solve the difference between the numbers. Then write an addition. Be careful.

a. from 3 to 5	**b.** from 1 to 5	**c.** from 2 to 7
_____ steps	_____ steps	_____ steps
$3 + \underline{\ 2\ } = 5$	$1 + \rule{1cm}{0.15mm} = 5$	$2 + \rule{1cm}{0.15mm} = 7$

6. Solve the difference between the numbers. Then write an addition. Be careful.

a. from 6 to 9: _____ steps 6 + _____ = 9	**b.** from 4 to 8: _____ steps _____ + _____ = _____
c. from 8 to 9: _____ steps _____ + _____ = _____	**d.** from 2 to 6: _____ steps _____ + _____ = _____

7. Who has more marbles? How many more? Write a "how many more" addition.
   You can also draw.

☐ Jill ☐ Sue **a.** Jill has 10 and Sue has 7. 7 + _____ = 10 __Jill__ has _3_ more.	☐ Ed ☐ Al **b.** Ed has 4 and Al has 7. _____ + _____ = _____ _____ has _____ more.
c. Veronica has 4 and Ann has 6. _____ + _____ = _____ _____ has _____ more.	**d.** Hannah has 9 and Michelle has 2. _____ + _____ = _____ _____ has _____ more.
e. Britney has 11 and Jeanne has 10. _____ + _____ = _____ _____ has _____ more.	**f.** Ben has 5 and Don has 10. _____ + _____ = _____ _____ has _____ more.

8. Solve the problems. Think carefully: Is it asking for the <u>total</u>?
   OR is it asking, "How many <u>more</u>?"

a. There are two cassette tapes on the table, and eight on the shelf.

How many tapes are there together?

How many more tapes are on the shelf?

b. There are five birds in the apple tree, and there are five more birds over in the oak tree. Four more flew into the oak tree.

How many birds are now in the oak tree?

How many more birds are in the oak tree than in the apple tree?

c. Brenda has 2 toy cars, Jason has 9, and Joe has 10.

How many more does Joe have than Jason?

How many more does Jason have than Brenda?

"How Many More" Problems and Subtraction

Nathalie needs ten safety-pins. She only has two.
<u>How many more</u> does she need?

$$ + \boxed{?} = 10$$

$$2 + \boxed{} = 10$$

You can use subtraction to solve a problem that asks "how many more", like the problem $2 + \underline{} = 10$.

Simply subtract the numbers: $10 - 2 = \underline{}$.

It gives you the answer to $2 + \underline{} = 10$. Why?

Because both problems are <u>in the same fact family</u>.

Use the Fact Family:

$2 + 8 = 10, \quad 10 - 2 = 8$

$8 + 2 = 10, \quad 10 - 8 = 2$

1. Fill in. The "how many more" problem has the same answer as the subtraction problem!

$+ \boxed{}$	$+ \boxed{}$	$+ \boxed{}$
a. $5 \quad + \underline{} = 7$	**b.** $3 \quad + \underline{} = 8$	**c.** $4 \quad + \underline{} = 9$
$7 - 5 = \underline{}$	$8 - 3 = \underline{}$	$9 - 4 = \underline{}$
$+ \boxed{}$	$+ \boxed{}$	$+ \boxed{}$
d. $5 \quad + \underline{} = 10$	**e.** $4 \quad + \underline{} = 7$	**f.** $5 \quad + \underline{} = 8$
$10 - 5 = \underline{}$	$7 - 4 = \underline{}$	$8 - 5 = \underline{}$
g. $3 + \underline{} = 10$	**h.** $2 + \underline{} = 9$	**i.** $1 + \underline{} = 7$
$10 - 3 = \underline{}$	$9 - 2 = \underline{}$	$7 - 1 = \underline{}$

2. Solve the subtraction problem *first*. (It is probably easier.) Then copy the answer to the "how many more" problem.

a. 2 + _____ = 8	**b.** 1 + _____ = 9	**c.** _____ + 3 = 10	**d.** _____ + 3 = 9
8 – 2 = _____	9 – 1 = _____	10 – 3 = _____	9 – 3 = _____

The "how many more" problem has the same answer as the subtraction problem!

3. Write a subtraction problem, using the same numbers, under each "how many more" problem. Solve the subtraction problem *first*.

a. 1 + _____ = 7 _____ – _____ = _____	**b.** 2 + _____ = 9 _____ – _____ = _____	**c.** 1 + _____ = 10 _____ – _____ = _____
d. _____ + 3 = 8 _____ – _____ = _____	**e.** _____ + 2 = 10 _____ – _____ = _____	**f.** _____ + 3 = 9 _____ – _____ = _____

4. Solve. Think: Do you already know the total? Or is the problem asking for the total? You can also draw a picture to help!

a. Mary ate two carrots. The rabbit ate six carrots. What was the total number of carrots eaten?	**b.** Baby put three blocks in a stack, and another four blocks in another stack. How many blocks did the baby use?
c. There were five lambs in the pen. Two more lambs went into the pen. How many lambs are now in the pen?	**d.** Lisa needs 8 dollars for a stuffed hippo. She has saved 4 dollars. How many more dollars does she need?

5. First write a "how many more" problem for each subtraction problem, using the same numbers. Then solve the easier problem. Copy the answer to the other problem.

a. $8 - 6 = $ ____ ____ $+$ ____ $=$ ____	**b.** $10 - 9 = $ ____ ____ $+$ ____ $=$ ____	**c.** $9 - 7 = $ ____ ____ $+$ ____ $=$ ____
d. $10 - 8 = $ ____ ____ $+$ ____ $=$ ____	**e.** $9 - 8 = $ ____ ____ $+$ ____ $=$ ____	**f.** $7 - 6 = $ ____ ____ $+$ ____ $=$ ____

6. Solve. Think: Do you already know the total? Or is the problem asking for the total? Write an addition or subtraction for each problem. You can also draw a picture to help!

a. Mom needs six cucumbers. She already has three. How many more does she need?	**b.** There were seven ducks on the pond. Three flew away. How many were left?
c. Jane wants to buy a teddy bear for 8 dollars. She has saved 6 dollars. How much more money does she need?	**d.** A book has ten pages. Jerry has read six pages. How many pages does he have left to read?

7. Play math teacher again. Bill joined Ann and Joe to work some more problems. Check their work and correct any mistakes that they made.

Ann:	**Joe:**	**Bill:**
a. $7 - \triangle{1} = 6$	**c.** $\triangle{2} - 4 = 2$	**e.** $\triangle{9} - 4 = 5$
b. $8 - \triangle{3} = 4$	**d.** $9 - \triangle{6} = 3$	**f.** $9 - \triangle{8} = 2$

8. These are the toys that Zach and Mary have.

a. How many dolls do the children have?

b. How many teddy bears?

c. How many other toys?

Zach and Mary want to make a graph of their toys. To complete the graph, draw one block for each toy. Draw them all the same size and lined up in a column, just like the ones for the dolls.

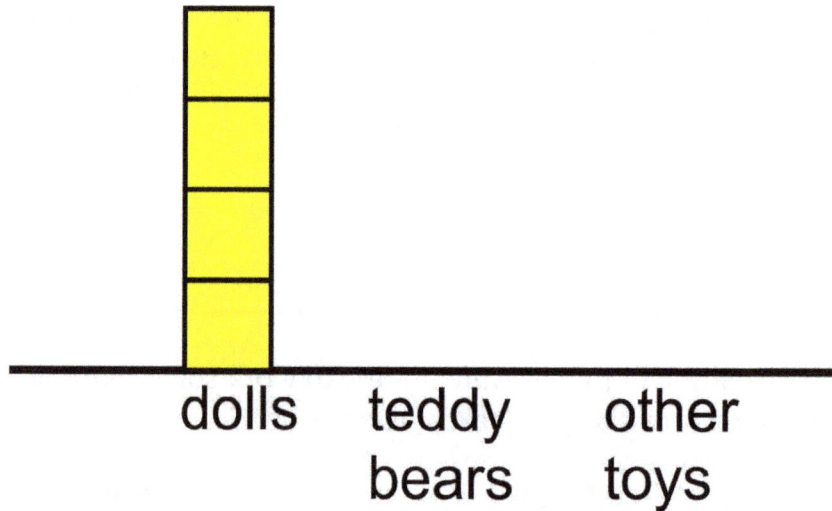

d. How many more teddy bears do they have than dolls?

e. How many more teddy bears do they have than other toys?

f. How many dolls and teddy bears do the children have in all?

Subtracting More Than One Number

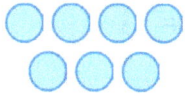

7 – 1 – 2 = ?

You have 7 balls. First you take away 1 ball, and then you take away 2 more balls. You will have 4 balls left. 7 – 1 – 2 = 4.

1. Subtract twice, taking away circles. You can cover the circles to help.

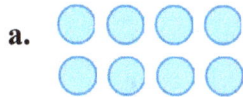

a.

8 – 2 – 3 = _____

8 – 5 – 2 = _____

8 – 1 – 3 = _____

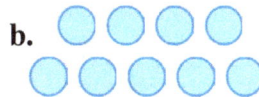

b.

9 – 3 – 1 = _____

9 – 4 – 2 = _____

9 – 2 – 5 = _____

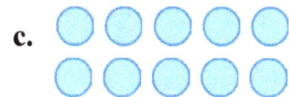

c.

10 – 5 – 3 = _____

10 – 6 – 2 = _____

10 – 1 – 4 = _____

2. Solve. You can draw pictures to help.

a. Mary had ten cookies. She gave two to her
brother and two to her sister.
How many does she have left?

b. Seven birds were in the tree. Three flew away.
After a while, one more flew away.
How many birds were left in the tree?

c. Eight cars were in the parking lot. Then three
cars left. After that two more cars left.
How many cars are there now?

d. Jack had $5. His mom gave him $1,
and his dad gave him $2.
How many dollars does Jack have now?

You can subtract two numbers one at a time:	OR you can subtract their total:
8 – 2 – 3	8 – 2 – 3
6 – 3 = 3	8 – 5 = 3
First take away 2. That leaves 6. Then, from 6, subtract 3. That leaves 3.	Check how much you need to subtract or take away *in total*. You need to subtract 2 and 3, or a total of 5. So, subtract 8 – 5 = 3.

3. Subtract by either method.

a.	b.	c.
$7 - 2 - 3 =$ _____	$9 - 7 - 1 =$ _____	$7 - 5 - 1 =$ _____
$9 - 2 - 6 =$ _____	$6 - 2 - 2 =$ _____	$10 - 6 - 1 =$ _____

4. Solve. Compare the two problems and their results.

a.	b.	c.
$10 - 3 - 2 =$ _____	$7 - 3 - 3 =$ _____	$9 - 6 - 1 =$ _____
$10 - 3 - 3 =$ _____	$7 - 4 - 3 =$ _____	$8 - 6 - 1 =$ _____

5. Match the subtraction problems to the correct pictures.

a. $8 - 2 - 2 - 2 - 2 = 0$

b. $8 - 4 - 4 = 0$

c. $6 - 2 - 2 - 2 = 0$

d. $6 - 3 - 3 = 0$

Puzzle Corner   Here are some problems with four numbers!

$9 - 3 - 2 - 1 =$ _____ $10 - 1 - 2 - 1 =$ _____ $8 - 4 - 1 - 2 =$ _____

46

Review

1. Write a fact family to match the picture.

____ + ____ = ____   ____ – ____ = ____

____ + ____ = ____   ____ – ____ = ____

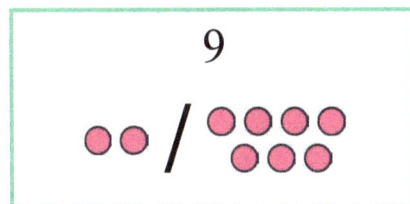

9

2. **a.** Write a subtraction that matches the addition $6 + 4 = 10$.   ____ – ____ = ____

  b. Write a subtraction to match the addition $5 + \underline{\quad} = 9$.   ____ – ____ = ____
     Solve both the addition and the matching subtraction.

3. **a.** There are 8 children playing in the yard. Two are girls.
     How many are boys?

  b. Kay has four marbles. Susan has two more marbles than Kay.
     Draw Kay's and Susan's marbles.

  c. Five sparrows and two robins are feeding on seeds. One more robin flies in.
     How many more sparrows than robins are there now?

4. Find the missing numbers.

a.	b.	c.	d.
$3 + \underline{\quad} = 4$	$6 - 3 = \underline{\quad}$	$10 - 0 = \underline{\quad}$	$8 - 2 = \underline{\quad}$
$1 + \underline{\quad} = 9$	$8 - 5 = \underline{\quad}$	$5 - 3 = \underline{\quad}$	$7 - 3 = \underline{\quad}$
$3 + \underline{\quad} = 10$	$7 - 6 = \underline{\quad}$	$6 - 6 = \underline{\quad}$	$10 - 1 = \underline{\quad}$
$2 + \underline{\quad} = 7$	$10 - 8 = \underline{\quad}$	$7 - 4 = \underline{\quad}$	$9 - 2 = \underline{\quad}$

Addition and Subtraction Facts with 4 and 5

Facts with 4	🦬🦬🦬🦬	4 + 0 = 4 0 + 4 = 4	4 − 4 = 0 4 − 0 = 4
	🦬🦬🦬 🦬	1 + 3 = 4 3 + 1 = 4	4 − 3 = 1 4 − 1 = 3
	🦬🦬 🦬🦬	2 + 2 = 4	4 − 2 = 2
Facts with 5	🦬🦬🦬🦬🦬	5 + 0 = 5 ____ + ____ = 5	5 − 5 = 0 5 − ____ = ____
	🦬🦬🦬🦬 🦬	4 + 1 = 5 1 + 4 = 5	5 − 4 = ____ 5 − ____ = ____
	🦬🦬🦬 🦬🦬	3 + 2 = 5 ____ + ____ = 5	5 − 3 = ____ 5 − ____ = ____

1. Find the missing numbers.

a.	b.	c.	d.
3 + _____ = 4	2 + _____ = 5	5 − 0 = _____	4 − 0 = _____
1 + _____ = 4	1 + _____ = 5	5 − 4 = _____	4 − 3 = _____
1 + _____ = 5	4 + _____ = 5	5 − 2 = _____	5 − 1 = _____
2 + _____ = 4	3 + _____ = 5	4 − 1 = _____	4 − 2 = _____

2. Color the square:

- yellow if the answer is 0.
- red if the answer is 1,
- blue if the answer is 2,
- green if the answer is 3,
- purple if the answer is 4,
- orange if the answer is 5.

$5 - 4$	$2 + 3$	$4 - 4$	$1 + 2$	$4 - 2$	$1 + 3$
$2 + 2$	$3 - 2$	$5 - 0$	$0 + 0$	$5 - 2$	$1 + 1$
$0 + 2$	$5 - 1$	$0 + 1$	$1 + 4$	$0 - 0$	$4 - 1$

3. Continue the patterns until the boxes are full.

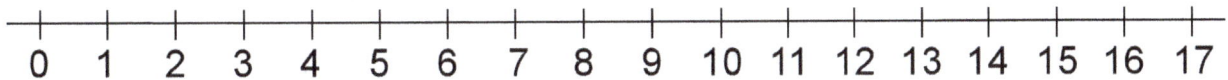

0 1 2 3 4 5 6 7 8 9 10 11 12 13 14 15 16 17

a.

$17 - 0 = \underline{\qquad}$

$17 - 1 = \underline{\qquad}$

$17 - 2 = \underline{\qquad}$

$17 - \underline{\quad} = \underline{\quad}$

$17 - \underline{\quad} = \underline{\quad}$

$\underline{\quad} - \underline{\quad} = \underline{\quad}$

$\underline{\quad} - \underline{\quad} = \underline{\quad}$

b.

$10 + \underline{\quad} = 10$

$10 + \underline{\quad} = 11$

$10 + \underline{\quad} = 12$

$10 + \underline{\quad} = \underline{\quad}$

$\underline{\quad} + \underline{\quad} = \underline{\quad}$

$\underline{\quad} + \underline{\quad} = \underline{\quad}$

$\underline{\quad} + \underline{\quad} = \underline{\quad}$

c.

$5 - 2 = \underline{\qquad}$

$6 - 2 = \underline{\qquad}$

$7 - 2 = \underline{\qquad}$

$\underline{\quad} - 2 = \underline{\quad}$

$\underline{\quad} - \underline{\quad} = \underline{\quad}$

$\underline{\quad} - \underline{\quad} = \underline{\quad}$

$\underline{\quad} - \underline{\quad} = \underline{\quad}$

Addition and Subtraction Facts with 6

1. Complete the fact families in which the sum is six. At the top, write the three numbers that you are using for the fact family.

6, 0, 6

$6 + 0 = 6$

_____ $+$ _____ $= 6$

$6 - 6 = 0$

$6 - 0 = 6$

_____, _____, 6

$5 + 1 = 6$

_____ $+$ _____ $= 6$

$6 - 5 = $ _____

$6 - $ _____ $= $ _____

_____, _____, 6

$4 + 2 = 6$

_____ $+$ _____ $= 6$

$6 - 4 = $ _____

$6 - $ _____ $= $ _____

_____, _____, 6

$3 + 3 = 6$

$6 - 3 = $ _____

2. Write the numbers that add up to 6. Memorize these!

$0 + $ _____ $= 6$ or   _____ $+ 0 = 6$

$1 + $ _____ $= 6$ or   _____ $+ 1 = 6$

$2 + $ _____ $= 6$ or   _____ $+ 2 = 6$

$3 + $ _____ $= 6$

3. Subtract.

a. $\begin{array}{r} 6 \\ -5 \\ \hline \end{array}$
b. $\begin{array}{r} 6 \\ -4 \\ \hline \end{array}$
c. $\begin{array}{r} 6 \\ -6 \\ \hline \end{array}$
d. $\begin{array}{r} 6 \\ -2 \\ \hline \end{array}$
e. $\begin{array}{r} 6 \\ -1 \\ \hline \end{array}$
f. $\begin{array}{r} 6 \\ -3 \\ \hline \end{array}$

4. Play the "6 Out" card game (see the introduction to this book).

5. Find the missing numbers.

a.	b.	c.	d.
4 + ____ = 6	5 + ____ = 6	____ + 2 = 6	6 − ____ = 2
3 + ____ = 6	0 + ____ = 6	____ + 1 = 6	6 − ____ = 5

6. For each "how many more" addition, write a subtraction using the same numbers so that the numbers in the triangles are the same.

a. 2 + △ = 5

__5__ − 2 = △3

b. 1 + △ = 6

____ − ____ = △

c. 4 + △ = 5

____ − ____ = △

d. 3 + △ = 8

__8__ − 3 = △

e. 5 + △ = 10

____ − ____ = △

f. 2 + △ = 7

____ − ____ = △

7. Write fact families so that the numbers in the triangles are the same.

a.   1 + △ = 6

△ + ____ = 6

____ − △ = ____

____ − ____ = △

b.   2 + 7 = △

____ + ____ = △

△ − ____ = ____

____ − ____ = ____

8. Solve the word problems. Think:

- Are you asked for the total? (2 + 4 = ?) OR
- Are you asked how many more? (2 + ? = 4) OR
- Are you asked how many are left? (4 − 2 = ?)

---

a. The black cat has four kittens and the white cat has three.
How many kittens do they have in total?

How many more kittens does the black cat have than the white cat?

---

b. John had ten crayons but now he only has two.
How many crayons are missing?

---

c. Mother found ten clothespins in one container and two in another.
How many clothespins were in the two containers?

How many more clothespins were in the first container
than in the second?

---

d. Jill has two eggs. She needs eight eggs to make some
cakes. How many more eggs does Jill need?

Her neighbor has three eggs. If the neighbor gives Jill
the three eggs she has, how many more does Jill still
need?

Addition and Subtraction Facts with 7

1. Complete the fact families with 7. At the top, write the three numbers you are using.

7, 0, 7

$7 + 0 = 7$

_____ + _____ = 7

$7 - $ _____ = _____

$7 - $ _____ = _____

_____, _____, 7

$6 + $ _____ = 7

_____ + _____ = 7

$7 - $ _____ = _____

_____ - _____ = _____

_____, _____, 7

$5 + $ _____ = 7

_____ + _____ = 7

$7 - $ _____ = _____

_____ - _____ = _____

_____, _____, 7

$4 + $ _____ = 7

_____ + _____ = 7

$7 - $ _____ = _____

_____ - _____ = _____

2. Write down the numbers that add up to 7 and memorize the addition facts!

$0 + $ _____ $ = 7$ or   _____ $ + 0 = 7$

$1 + $ _____ $ = 7$ or   _____ $ + 1 = 7$

$2 + $ _____ $ = 7$ or   _____ $ + 2 = 7$

$3 + $ _____ $ = 7$ or   _____ $ + 3 = 7$

3. Play the "7 Out" card game (see the introduction to this book).

4. Subtract.

a.	b.	c.	d.	e.	f.
$7 - 5$	$7 - 4$	$7 - 6$	$7 - 2$	$7 - 1$	$7 - 3$

5. Fill in. Then draw a line between the facts that are from the same fact family.

$5 +$ _____ $= 7$

$7 - 0 =$ _____

$7 - 3 =$ _____

$7 -$ _____ $= 1$

$7 -$ _____ $= 4$

$6 +$ _____ $= 7$

$7 - 2 =$ _____

$0 +$ _____ $= 7$

$7 -$ _____ $= 6$

_____ $+ 4 = 7$

$7 -$ _____ $= 7$

$7 -$ _____ $= 2$

6. Solve.

a. Luis has 4 pencils and Jeremy has 6.
How many more pencils does Jeremy have than Luis?

How many pencils do the two boys have in total?

b. Maria found two socks in the hamper, five socks in her basket,
and one sock on the floor. How many socks did she find?

Puzzle Corner   Figure out how to fill in the rest of this subtraction table.

−	12	11	10	9	8	7	6	5	4	3
1	11					6				
2		9					4			1

54

Addition and Subtraction Facts with 8

1. Complete the fact families which have the sum of eight.

8, _____, 8	_____, _____, 8	_____, _____, 8
8 + 0 = 8	7 + 1 = 8	___ + ___ = ___
___ + ___ = 8	___ + ___ = 8	___ + ___ = ___
8 – ___ = ___	8 – ___ = ___	___ – ___ = ___
___ – ___ = ___	___ – ___ = ___	___ – ___ = ___

_____, _____, 8	_____, _____, 8
___ + ___ = 8	___ + ___ = ___
___ + ___ = 8	
___ – ___ = ___	___ – ___ = ___
___ – ___ = ___	

2. Play the "8 Out" card game (see the introduction to this book).

3. Write the addition facts with 8 and memorize them!

$0 +$ _____ $= 8$ or   _____ $+ \ 0 \ = 8$

$1 +$ _____ $= 8$ or   _____ $+ \ 1 \ = 8$

$2 +$ _____ $= 8$ or   _____ $+$ _____ $= 8$ $4 +$ _____ $= 8$

$3 +$ _____ $= 8$ or   _____ $+$ _____ $= 8$

4. Find the missing numbers.

a.	b.	c.	d.
$8 - 3 =$ _____	$5 +$ _____ $= 8$	$4 +$ _____ $= 8$	$8 - 7 =$ _____
$8 - 1 =$ _____	$1 +$ _____ $= 8$	$6 +$ _____ $= 8$	$8 - 0 =$ _____
$8 - 2 =$ _____	$2 +$ _____ $= 8$	$7 +$ _____ $= 8$	$8 - 4 =$ _____

First subtract $6 - 2$ to get 4.
Write 4 in the shaded box below.

$6 - 2$ **?** 5

$\downarrow$ $\downarrow$

Then compare 4 and 5.
Since four is less than five, write "$<$".

4 $<$ 5

5. First subtract. Write the answer in the shaded box below.

a. $5 - 2$ **?** 4	b. $7 - 4$ **?** 5	c. $8 - 1$ **?** 7	d. $6 - 3$ **?** 2
$\downarrow$ $\downarrow$	$\downarrow$ $\downarrow$	$\downarrow$ $\downarrow$	$\downarrow$ $\downarrow$
$\square$ 4	$\square$ 5	$\square$	$\square$

First find $4 + 2$ (on the left side) and $8 - 3$ (on the right side).

Write the answers below. Then compare. Which is greater? Six is greater. So write ">" in the box.

$$4 + 2 \quad ? \quad 8 - 3$$

$$6 \quad \boxed{>} \quad 5$$

6. First add and subtract. Write the answer in the box below. Then compare, and write $<$, $>$ or $=$.

a. $5 - 2 \quad ? \quad 4 - 2$

b. $8 - 1 \quad ? \quad 7 - 1$

c. $8 - 6 \quad ? \quad 8 - 5$

d. $6 + 2 \quad ? \quad 7 + 2$

e. $7 - 1 \quad ? \quad 7 - 2$

f. $4 + 4 \quad ? \quad 7 - 5$

g. $1 - 1 \quad ? \quad 3 - 2$

h. $3 + 10 \quad ? \quad 10$

i. $7 \quad ? \quad 4 + 2$

j. $8 - 1 \quad ? \quad 4 - 2$

k. $7 - 2 \quad ? \quad 6 - 1$

l. $9 - 0 \quad ? \quad 7 + 2$

7. Fill in the missing numbers. Then draw a line between the facts that are from the same fact family.

6 + ____ = 8	8 – 4 = ____	____ + 5 = 8
8 – 0 = ____	8 – ____ = 3	8 – ____ = 6
8 – 3 = ____	7 + ____ = 8	____ + 4 = 8
____ + 1 = 8	8 – 2 = ____	8 – ____ = 8
4 + ____ = 8	0 + ____ = 8	8 – 1 = ____

8. Solve the word problems. Drawing can help you.

a. Jack has 10 cars, Bill has 7, and Ed has 4.

How many more cars than Ed does Bill have?

How many more cars than Ed does Jack have?

How many more cars than Bill does Jack have?

b. Mary has saved seven dollars. She would like to buy a puzzle for five dollars and a game for three dollars.
Can she buy both things?

If she can, how much money does she have left over?

If she can't, how much more money would she need to save?

Addition and Subtraction Facts with 9

1. Write the fact families where the sum is 9.

9, 0, 9

9 + _____ = 9

_____ + _____ = 9

9 − _____ = _____

_____ − _____ = _____

_____, _____, 9

_____ + _____ = _____

_____ + _____ = _____

_____ − _____ = _____

_____ − _____ = _____

_____, _____, 9

_____ + _____ = _____

_____ + _____ = _____

_____ − _____ = _____

_____ − _____ = _____

_____, _____, 9

_____ + _____ = _____

_____ + _____ = _____

_____ − _____ = _____

_____ − _____ = _____

_____, _____, 9

_____ + _____ = _____

_____ + _____ = _____

_____ − _____ = _____

_____ − _____ = _____

2. Play the "9 Out" card game (see the introduction to this book).

3. Write the addition
   facts with 9 and
   memorize them!

$0 +$ _____ $= 9$ or _____ $+ \ 0 \ = 9$

$1 +$ _____ $= 9$ or _____ $+ \ 1 \ = 9$

$2 +$ _____ $= 9$ or _____ $+$ _____ $= 9$

$3 +$ _____ $= 9$ or _____ $+$ _____ $= 9$

$4 +$ _____ $= 9$ or _____ $+$ _____ $= 9$

4. Find the missing numbers.

a.	b.	c.	d.
$9 - 5 =$ _____	$2 +$ _____ $= 9$	_____ $+ 1 = 9$	$9 -$ _____ $= 1$
$9 - 3 =$ _____	$1 +$ _____ $= 9$	_____ $+ 3 = 9$	$9 -$ _____ $= 0$
$9 - 6 =$ _____	$7 +$ _____ $= 9$	_____ $+ 5 = 9$	$9 -$ _____ $= 2$
$9 - 8 =$ _____	$8 +$ _____ $= 9$	_____ $+ 7 = 9$	$9 -$ _____ $= 4$

5. Fill in the missing numbers. Then draw lines to connect the facts that
   belong to the same fact family.

$7 +$ _____ $= 9$

$9 - 3 =$ _____

$9 - 1 =$ _____

$9 -$ _____ $= 9$

$4 +$ _____ $= 9$

$0 +$ _____ $= 9$

$9 -$ _____ $= 3$

$5 +$ _____ $= 9$

$9 - 2 =$ _____

$9 -$ _____ $= 1$

_____ $+ 8 = 9$

$9 -$ _____ $= 6$

_____ $+ 2 = 9$

$9 -$ _____ $= 4$

_____ $+ 0 = 9$

6. First add or subtract. Do not write answers in the box—just solve them in your head! Then compare and write $<$, $>$ or $=$ in the box.

a. 8 ☐ $10 - 3$ b. 9 ☐ $9 + 3$ c. $8 - 6$ ☐ $6 + 3$

d. $6 + 2$ ☐ $8 + 2$ e. $10 - 1$ ☐ 10 f. $8 - 4$ ☐ $8 - 5$

g. $5 - 2$ ☐ $4 - 2$ h. $8 + 0$ ☐ $8 - 0$ i. $9 - 1$ ☐ $9 + 1$

7. Subtract.

a. $\begin{array}{r} 9 \\ -5 \\ \hline \end{array}$ b. $\begin{array}{r} 9 \\ -4 \\ \hline \end{array}$ c. $\begin{array}{r} 9 \\ -6 \\ \hline \end{array}$ d. $\begin{array}{r} 8 \\ -2 \\ \hline \end{array}$ e. $\begin{array}{r} 9 \\ -2 \\ \hline \end{array}$ f. $\begin{array}{r} 8 \\ -3 \\ \hline \end{array}$

8. If the answer is 6 or 7, color its box blue. If the answer is 8 or 9, color its box red. Color the rest of the boxes yellow.

$9 - 3$	$4 + 6$	$9 - 0$	$4 + 6$	$8 - 1$
$2 + 5$	$9 - 5$	$4 + 4$	$4 - 2$	$5 + 1$
$9 - 2$	$3 + 7$	$10 - 2$	$10 + 0$	$7 - 1$
$4 + 2$	$7 - 3$	$6 + 3$	$3 - 1$	$3 + 3$
$6 - 0$	$1 + 1$	$8 - 0$	$3 + 2$	$10 - 4$
$3 + 4$	$8 - 3$	$2 + 7$	$7 - 6$	$7 + 0$
$1 + 6$	$2 + 8$	$10 - 1$	$2 + 2$	$7 - 0$

Addition and Subtraction Facts with 10

1. Complete the fact families which have the sum of ten.

10, 0, 10	9, 1, 10	____, ____, 10
$10 + 0 = 10$	$9 + \underline{\quad} = 10$	$8 + \underline{\quad} = 10$
$\underline{\quad} + \underline{\quad} = 10$	$\underline{\quad} + \underline{\quad} = 10$	$\underline{\quad} + \underline{\quad} = 10$
$10 - 0 = \underline{\quad}$	$10 - 9 = \underline{\quad}$	$10 - \underline{\quad} = \underline{\quad}$
$10 - \underline{\quad} = \underline{\quad}$	$10 - \underline{\quad} = \underline{\quad}$	$10 - \underline{\quad} = \underline{\quad}$

____, ____, 10	____, ____, 10	____, ____, 10
$7 + \underline{\quad} = 10$	$\underline{\quad} + \underline{\quad} = \underline{\quad}$	$\underline{\quad} + \underline{\quad} = \underline{\quad}$
$\underline{\quad} + \underline{\quad} = 10$	$\underline{\quad} + \underline{\quad} = \underline{\quad}$	$\underline{\quad} + \underline{\quad} = \underline{\quad}$
$10 - \underline{\quad} = \underline{\quad}$	$\underline{\quad} - \underline{\quad} = \underline{\quad}$	$\underline{\quad} - \underline{\quad} = \underline{\quad}$
$10 - \underline{\quad} = \underline{\quad}$	$\underline{\quad} - \underline{\quad} = \underline{\quad}$	$\underline{\quad} - \underline{\quad} = \underline{\quad}$

2. Play the "10 Out" card game (see the introduction to this book).

3. Write the addition facts with 10 and memorize them!

$0 + \underline{\hspace{1cm}} = 10$ or $\underline{\hspace{1cm}} + 0 = 10$

$1 + \underline{\hspace{1cm}} = 10$ or $\underline{\hspace{1cm}} + 1 = 10$

$2 + \underline{\hspace{1cm}} = 10$ or $\underline{\hspace{1cm}} + \underline{\hspace{1cm}} = 10$

$3 + \underline{\hspace{1cm}} = 10$ or $\underline{\hspace{1cm}} + \underline{\hspace{1cm}} = 10$

$4 + \underline{\hspace{1cm}} = 10$ or $\underline{\hspace{1cm}} + \underline{\hspace{1cm}} = 10$

$5 + \underline{\hspace{1cm}} = 10$

4. Find the missing numbers.

a.	b.	c.
$10 - 3 = \underline{\hspace{1cm}}$	$0 + \underline{\hspace{1cm}} = 10$	$10 - \underline{\hspace{1cm}} = 7$
$10 - 6 = \underline{\hspace{1cm}}$	$7 + \underline{\hspace{1cm}} = 10$	$10 - \underline{\hspace{1cm}} = 2$
$10 - 2 = \underline{\hspace{1cm}}$	$9 + \underline{\hspace{1cm}} = 10$	$10 - \underline{\hspace{1cm}} = 4$

5. Fill in. Then draw lines to connect the facts that are from the same fact family.

$6 + \underline{\hspace{1cm}} = 10$	$0 + \underline{\hspace{1cm}} = 10$	$\underline{\hspace{1cm}} + 5 = 10$
$10 - 5 = \underline{\hspace{1cm}}$	$8 + \underline{\hspace{1cm}} = 10$	$10 - \underline{\hspace{1cm}} = 6$
$10 - 1 = \underline{\hspace{1cm}}$	$10 - \underline{\hspace{1cm}} = 3$	$\underline{\hspace{1cm}} + 2 = 10$
$10 - 2 = \underline{\hspace{1cm}}$	$5 + \underline{\hspace{1cm}} = 10$	$10 - \underline{\hspace{1cm}} = 7$
$10 - \underline{\hspace{1cm}} = 10$	$10 - 4 = \underline{\hspace{1cm}}$	$\underline{\hspace{1cm}} + 10 = 10$
$7 + \underline{\hspace{1cm}} = 10$	$9 + \underline{\hspace{1cm}} = 10$	$10 - 9 = \underline{\hspace{1cm}}$

6. Write an addition sentence and a subtraction sentence for these word problems.
Remember, you can always draw a picture of the situation to help you!

a. Sarah has six coins in her piggy bank. Elisa has two coins in hers.

Today Elisa found three coins on the ground.

Now who has more coins?

How many more?

b. Dad had one box of nails at home, and then he bought six more boxes of nails. The next day he gave one box to a neighbor.

How many boxes of nails does Dad have now?

c. You have six dollars and I have three. Can we share the cost of a meal that costs $8?

If so, is there any money left over, and how much?

If not, how much more money would we need?

7. More missing numbers!

a. $10 - \boxed{} = 5$	**b.** $3 + \boxed{} = 10$	**c.** $10 - \boxed{} = 6$
$10 - \boxed{} = 7$	$4 + \boxed{} = 10$	$10 - \boxed{} = 7$
$10 - \boxed{} = 2$	$1 + \boxed{} = 10$	$10 - \boxed{} = 9$

8. Several girls counted how many coins they had. Make a bar graph. This means you need to draw the bars. Just draw rectangles so that they reach the given numbers.

Girl	Coins
Lisa	9
Maggie	11
Lily	6
Susana	8

Coin Count

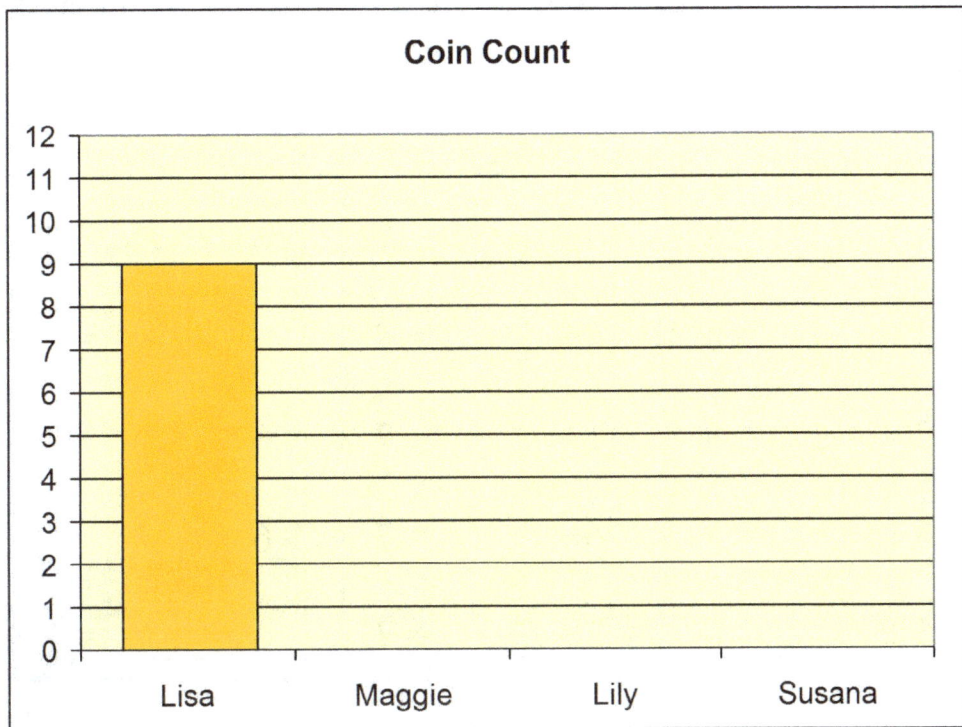

9. Make up two "how many more" questions about the bar graph. Ask a friend the questions. Check your friend's answers!

Puzzle Corner Let's let $\triangle$ represent one number, and $\square$ represent another number. They're different in each case (**a**, **b**, and **c**). For each case, find what the numbers are. Just guess and check!

a.

$$\square + \triangle = 10$$

$$\square - \triangle = 2$$

b.

$$\square + \triangle = 10$$

$$\square - \triangle = 6$$

c.

$$\square + \triangle = 10$$

$$\square - \triangle = 0$$

Review — Facts with 6, 7, and 8

1. Practice addition and subtraction facts with 6, 7, and 8.

a.	**b.**	**c.**	**d.**
$0 + \underline{\hspace{1cm}} = 8$	$3 + \underline{\hspace{1cm}} = 7$	$6 - \underline{\hspace{1cm}} = 2$	$7 - \underline{\hspace{1cm}} = 2$
$3 + \underline{\hspace{1cm}} = 8$	$5 + \underline{\hspace{1cm}} = 7$	$6 - \underline{\hspace{1cm}} = 5$	$8 - \underline{\hspace{1cm}} = 3$
$2 + \underline{\hspace{1cm}} = 8$	$1 + \underline{\hspace{1cm}} = 7$	$6 - \underline{\hspace{1cm}} = 3$	$6 - \underline{\hspace{1cm}} = 1$
$6 + \underline{\hspace{1cm}} = 8$	$6 + \underline{\hspace{1cm}} = 7$	$6 - \underline{\hspace{1cm}} = 4$	$8 - \underline{\hspace{1cm}} = 4$
$5 + \underline{\hspace{1cm}} = 8$	$2 + \underline{\hspace{1cm}} = 7$	$6 - \underline{\hspace{1cm}} = 1$	$7 - \underline{\hspace{1cm}} = 4$

2. First add and subtract. Write the answers below. Then compare, and write $<$, $>$ or $=$.

a. $8 - 2$ **?** $7 - 3$	**b.** $10 - 7$ **?** $9 - 6$	**c.** $7 - 6$ **?** $4 - 2$
$\downarrow \quad \square \quad \downarrow$	$\downarrow \quad \square \quad \downarrow$	$\downarrow \quad \square \quad \downarrow$
d. $4 + 2 \ \square \ 9 - 8$	**e.** $10 - 4 \ \square \ 7 - 4$	**f.** $3 + 4 \ \square \ 7 - 1$

3. Solve.

a. Luisa and Caleb were playing a game. Luisa had 9 game pieces and Caleb had 4. How many more game pieces did Luisa have than Caleb?

b. Luisa gave one game piece to Caleb. Now who has more game pieces?

How many more?

4. Complete. Then draw lines to connect the facts from the same fact family.

_____ − 5 = 1	7 − 5 = _____	8 − 3 = _____
2 + _____ = 7	_____ + 2 = 6	5 + _____ = 7
8 − _____ = 3	6 − 1 = _____	1 + 5 = _____
_____ + 2 = 8	5 + _____ = 8	8 − 6 = _____
6 − 4 = _____	8 − _____ = 6	2 + 4 = _____

5. Complete. Then draw a line to connect the facts from the same fact family.

3 + _____ = 7	_____ + 6 = 7	8 − _____ = 7
6 − _____ = 3	_____ − 7 = 1	1 + 6 = _____
_____ + 1 = 8	3 + 3 = _____	3 + _____ = 6
_____ − 4 = 4	4 + _____ = 7	8 − 4 = _____
7 − 1 = _____	8 − _____ = 4	7 − _____ = 4

Puzzle Corner

What numbers can go into the squares? All the numbers are less than 10. Guess and check!

	+		= 8
−		−	
	+		= 6
‖		‖	
0		2	

	+		= 7
+		−	
	−		= 4
‖		‖	
8		3	

67

Review — Facts with 9 and 10

1. Practice addition and subtraction facts with 9 and 10.

a.

$4 + \underline{\hspace{1cm}} = 9$

$1 + \underline{\hspace{1cm}} = 9$

$6 + \underline{\hspace{1cm}} = 9$

$2 + \underline{\hspace{1cm}} = 9$

b.

$5 + \underline{\hspace{1cm}} = 10$

$2 + \underline{\hspace{1cm}} = 10$

$3 + \underline{\hspace{1cm}} = 10$

$4 + \underline{\hspace{1cm}} = 10$

c.

$10 - \underline{\hspace{1cm}} = 1$

$10 - \underline{\hspace{1cm}} = 7$

$10 - \underline{\hspace{1cm}} = 5$

$10 - \underline{\hspace{1cm}} = 8$

d.

$9 - \underline{\hspace{1cm}} = 2$

$9 - \underline{\hspace{1cm}} = 6$

$9 - \underline{\hspace{1cm}} = 8$

$9 - \underline{\hspace{1cm}} = 5$

2. Match the addition problems to the right pictures and solve them.

a. $2 + 3 + 3 = \underline{\hspace{2cm}}$

b. $3 + 2 + 2 = \underline{\hspace{2cm}}$

c. $1 + 2 + 2 = \underline{\hspace{2cm}}$

d. $3 + 4 + 2 = \underline{\hspace{2cm}}$

e. $3 + 3 + 3 = \underline{\hspace{2cm}}$

f. $2 + 2 + 2 = \underline{\hspace{2cm}}$

3. Time to play teacher again! Ann, Joe, and Bill worked some math problems. Check their work, and correct any that are wrong.

Ann:

a. $5 - 0 = \boxed{5}$

b. $10 - \boxed{3} = 6$

Joe:

c. $9 - 4 = \boxed{6}$

d. $6 - \boxed{4} = 2$

Bill:

e. $7 - \boxed{5} = 3$

f. $\boxed{8} - 6 = 2$

4. Fill in the missing numbers. Draw lines to connect the facts that belong
   to the same fact family.

$9 - \underline{\hspace{1cm}} = 7$

$9 - \underline{\hspace{1cm}} = 6$

$9 - 1 = \underline{\hspace{1cm}}$

$9 - \underline{\hspace{1cm}} = 4$

$\underline{\hspace{1cm}} + 2 = 9$

$8 + \underline{\hspace{1cm}} = 9$

$\underline{\hspace{1cm}} + 5 = 9$

$3 + \underline{\hspace{1cm}} = 9$

$9 - \underline{\hspace{1cm}} = 5$

$9 - 6 = \underline{\hspace{1cm}}$

$9 - \underline{\hspace{1cm}} = 2$

$\underline{\hspace{1cm}} + 8 = 9$

5. **a.** Draw a line to connect each pair of
   numbers that add up to 9.
   Which number is left by itself?

```
0        7      2
    2       8      5
1      6
  9        4   3   4
      5
  1        3   8
7        6        9
```

   b. Draw a line to connect each pair of
   numbers that add up to 10.
   Which number is left by itself?

```
3        7   10
    2       8      9
1      6
  9        4   0   2
      5
  1        3   8
7        6        5
```

6. Fill in the missing numbers. Draw a line to connect the facts that are from
   the same fact family.

$10 - \underline{\hspace{1cm}} = 8$

$10 - \underline{\hspace{1cm}} = 5$

$10 - \underline{\hspace{1cm}} = 1$

$10 - 3 = \underline{\hspace{1cm}}$

$10 - 6 = \underline{\hspace{1cm}}$

$\underline{\hspace{1cm}} + 9 = 10$

$4 + \underline{\hspace{1cm}} = 10$

$5 + \underline{\hspace{1cm}} = 10$

$2 + \underline{\hspace{1cm}} = 10$

$\underline{\hspace{1cm}} + 7 = 10$

$\underline{\hspace{1cm}} + 1 = 10$

$10 - 5 = \underline{\hspace{1cm}}$

$10 - 4 = \underline{\hspace{1cm}}$

$\underline{\hspace{1cm}} + 3 = 10$

$10 - \underline{\hspace{1cm}} = 8$

7. Solve.

a. Millie has two boxes of crayons. Ken has seven boxes.
How many more boxes does Ken have?

b. Mike has three yellow toy cars, four blue ones, and
three red ones. How many toy cars does he have in all?

c. There were four birds in a tree. Four more flew in.
How many birds are there now?

Look! Five of them just flew away!
How many birds are there now?

d. Elisa knows she has ten crayons. She can only find four.
How many are missing?

e. A ten-piece puzzle has two pieces missing.
How many pieces are there now?

Answer Key

Subtraction is "Taking Away", pp. 9-11

Page 9

1.

a. $10 - 4 = 6$	b. $6 - 6 = 0$	c. $8 - 2 = 6$
d. $8 - 7 = 1$	e. $8 - 6 = 2$	f. $7 - 3 = 4$

Page 10

1. (continued)

g. $9 - 3 = 6$	h. $6 - 2 = 4$	i. $10 - 7 = 3$
j. $9 - 5 = 4$	k. $8 - 4 = 4$	l. $10 - 5 = 5$
m. $6 - 3 = 3$	n. $6 - 5 = 1$	o. $5 - 5 = 0$

2. a. 4   b. 3   c. 2   d. 3   e. 3   f. 4   g. 6   h. 6   i. 5   j. 6   k. 4   l. 4

Page 11

3. a. 5   b. 4   c. 3   d. 5   e. 9   f. 2   g. 3   h. 4   i. 6   j. 4   k. 4   l. 2

4. a. $8 - 4 = 4$ b. $7 - 4 = 3$
   c. $7 - 1 = 6$ d. $9 - 6 = 3$
   e. $9 - 7 = 2$ f. $10 - 2 = 8$

Count Down to Subtract, pp. 12-15

Page 12

1.

a. $7 - 2 = \underline{\quad 5 \quad}$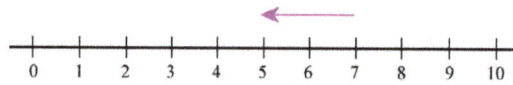

b. $8 - 4 = \underline{\quad 4 \quad}$

c. $6 - 5 = \underline{\quad 1 \quad}$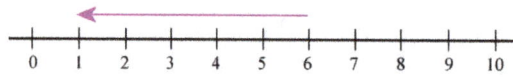

d. $9 - 3 = \underline{\quad 6 \quad}$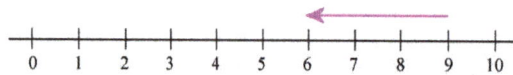

e. $10 - 3 = \underline{\quad 7 \quad}$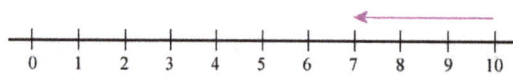

Page 13

2. a. $6 - 4 = 2$ b. $4 - 1 = 3$ c. $7 - 5 = 2$ d. $9 - 4 = 5$

3.

a. $10 - 5 = \underline{\quad 5 \quad}$

b. $7 - 6 = \underline{\quad 1 \quad}$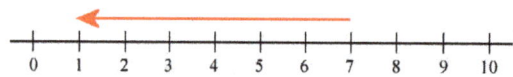

c. $4 - 4 = \underline{\quad 0 \quad}$

d. $8 - 4 = \underline{\quad 4 \quad}$

e. $10 - 1 = \underline{\quad 9 \quad}$

f. $7 - 5 = \underline{\quad 2 \quad}$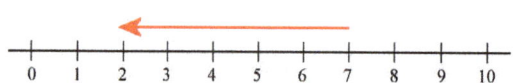

Page 14

4. a. <u>4</u>, 5, <u>6</u>   b. <u>1</u>, 2 , <u>3</u>   c. <u>7</u>, 8, <u>9</u>

   d. <u>5</u>, 6 , <u>7</u>   e. <u>3</u>, 4 , <u>5</u>   f. <u>8</u>, 9 , <u>10</u>

5. a. <u>5</u>, <u>6</u>, 7   b. <u>2</u>, <u>3</u>, 4   c. <u>8</u>, <u>9</u>, 10

   d. <u>4</u>, <u>5</u>, 6   e. <u>0</u>, <u>1</u>, 2   f. <u>6</u>, <u>7</u>, 8

6. a. 5, 7   b. 4, 6   c. 8, 2   d. 3, 9

7. a. 4, 3   b. 7, 6   c. 5, 4   d. 8, 7

8. All of these situations are of the type "taking away" or "going away" etc.
   a. $7 - 3 = 4$ There are 4 birds left in the tree.
   b. $10 - 4 = 6$ There are still six silver plates in the cupboard.
   c. $9 - 4 = 5$ Five girls kept jumping rope.
   d. $10 - 5 = 5$ Josh left five of his toy cars at home.
   e. $8 - 3 = 5$ Fanny still has five puzzles to play with.
   f. $6 - 6 = 0$ Tina does not have any flowers left.

9. a. 12, 10   b. 15, 13   c. 8, 7   d. 15, 14   e. 11, 10   f. 11, 10

Subtraction and Addition in the Same Picture, pp. 16-19

Page 16

1.

a. $3 + 4 = 7$ $7 - 4 = 3$ or $7 - 3 = 4$	b. $4 + 2 = 6$ $6 - 2 = 4$ or $6 - 4 = 2$
c. $1 + 4 = 5$ $5 - 4 = 1$ or $5 - 1 = 4$	d. $5 + 1 = 6$ $6 - 1 = 5$ or $6 - 5 = 1$
e. $5 + 3 = 8$ $8 - 3 = 5$ or $8 - 5 = 3$	f. $3 + 3 = 6$ $6 - 3 = 3$

Page 17

2.

a.	$5 + 4 = 9$ $9 - 4 = 5$ or $9 - 5 = 4$	b.	$3 + 6 = 9$ $9 - 6 = 3$ or $9 - 3 = 6$
c.	$5 + 5 = 10$ $10 - 5 = 5$	d.	$6 + 6 = 12$ $12 - 6 = 6$

3.

○○○○○○○● a. $7 + 1 = 8$ $8 - 1 = 7$	○○○○○○●●● b. $6 + 3 = 9$ $9 - 3 = 6$
○○●●● c. $2 + 3 = 5$ $5 - 3 = 2$	○○●●●●● d. $2 + 5 = 7$ $7 - 2 = 5$
○○○○○○○●●●● e. $7 + 4 = 11$ $11 - 4 = 7$	○○○●●● f. $3 + 3 = 6$ $6 - 3 = 3$

Page 18

4. a. $10 - 4 = 6$ b. $8 - 5 = 3$ c. $6 - 2 = 4$ d. $5 - 4 = 1$
   e. $5 - 1 = 4$ f. $6 - 3 = 3$ g. $11 - 7 = 4$ h. $11 - 6 = 5$

5. Either subtraction sentence could be correct, depending on which circles the student colored.

○○○○○○○○● a. $9 + 1 = 10$ $10 - 9 = 1$ or $10 - 1 = 9$	○○○○○○○●● b. $7 + 2 = 9$ $9 - 7 = 2$ or $9 - 2 = 7$
○○○○○○○○○○ ●●●● c. $10 + 4 = 14$ $14 - 10 = 4$ or $14 - 4 = 10$	○○○○○○○○○○ ●● d. $10 + 2 = 12$ $12 - 10 = 2$ or $12 - 2 = 10$

Subtraction and Addition in the Same Picture, cont.

Page 19

6.

○○○○○ ⊘⊘⊘⊘ a. 9 – 4 = 5 5 + 4 = 9	○○○○○⊘⊘⊘⊘⊘ b. 10 – 5 = 5 5 + 5 = 10
○○○⊘⊘⊘⊘⊘ c. 8 – 5 = 3 3 + 5 = 8	○○○○⊘⊘⊘⊘ d. 8 – 4 = 4 4 + 4 = 8
○○○⊘⊘⊘⊘ e. 7 – 4 = 3 4 + 3 = 7	○⊘⊘⊘⊘⊘⊘⊘⊘ f. 9 – 8 = 1 1 + 8 = 9

Puzzle Corner.

3 $>$ 3 – 1	6 + 5 $>$ 6	10 $>$ 10 – 1
9 – 7 $>$ 8 – 7	6 – 4 $<$ 2 + 3	8 – 5 $<$ 5 + 3
5 + 2 $<$ 8 + 2	10 – 1 $>$ 10 – 3	7 – 4 $=$ 8 – 5
10 – 2 $>$ 8 – 2	10 + 0 $=$ 10 – 0	8 – 1 $<$ 8 + 1

When Can You Subtract?, pp. 20-23

Page 20

1.

4 – 0 = 4 4 – 1 = 3 4 – 2 = 2 4 – 3 = 1 4 – 4 = 0	6 – 0 = 6 6 – 1 = 5 6 – 2 = 4 6 – 3 = 3 6 – 4 = 2 6 – 5 = 1 6 – 6 = 0	5 – 0 = 5 5 – 1 = 4 5 – 2 = 3 5 – 3 = 2 5 – 4 = 1 5 – 5 = 0

You cannot do a subtraction problem in whole numbers (0, 1, 2, 3, …) when the second number is **bigger / larger / greater** than the first number.

Page 21

2.

a.	b.	c.	d.
7 – 1 = 6 7 – 2 = 5 7 – 3 = 4	9 – 1 = 8 9 – 2 = 7 9 – 3 = 6	10 – 1 = 9 10 – 2 = 8 10 – 3 = 7	12 – 1 = 11 12 – 2 = 10 12 – 3 = 9

Page 21

3.

a.	b.	c.
$7 - 0 = 7$	$10 - 5 = 5$	$8 - 2 = 6$
$7 - 1 = 6$	$9 - 5 = 4$	$7 - 2 = 5$
$7 - 2 = 5$	$8 - 5 = 3$	$6 - 2 = 4$
$7 - 3 = 4$	$7 - 5 = 2$	$5 - 2 = 3$
$7 - 4 = 3$	$6 - 5 = 1$	$4 - 2 = 2$
$7 - 5 = 2$	$5 - 5 = 0$	$3 - 2 = 1$
$7 - 6 = 1$		$2 - 2 = 0$
$7 - 7 = 0$		

Page 22

4.

$4 - 0$	$7 - 7$	~~$5 - 6$~~	~~$3 - 6$~~	$4 - 4$	~~$3 - 10$~~
~~$4 - 5$~~	~~$7 - 9$~~	$10 - 1$	~~$3 - 4$~~	~~$2 - 4$~~	$4 - 3$

5. a. false   b. true   c. false   d. false   e. false   f. true.

6. You can draw pictures (rough sketches) to help the child with the actual math. This would mean drawing individual dollar bills. For example, in (a), you could draw three rectangles to be three dollar bills.

a. Jennie cannot buy the doll. She needs two dollars more.   b. Jessie can buy the ball. He will have three dollars left.
c. Lola can buy the Lego set and have three dollars left.   d. Marvin cannot buy the book. He needs one dollar more.
e. Jack cannot buy the set. He needs two dollars more.   f. Mary can buy the car. She will have five dollars left.
g. Faye can buy the game and have eight dollars left.   h. He cannot buy the game. He needs five dollars more.

Page 23

7.

a. $10 - 3 = 7$ b. $8 - 7 = 1$ c. $6 - 5 = 1$ d. $8 - 6 = 2$ e. $8 - 0 = 8$

f. $7 - 7 = 0$ g. $7 - 6 = 1$ h. $6 - 6 = 0$ i. $6 - 1 = 5$ j. $9 - 4 = 5$

k. $10 - 8 = 2$ l. $4 - 0 = 4$ m. $6 - 4 = 2$ n. $7 - 2 = 5$ o. $9 - 3 = 6$

8.

$10 - 6$	$6 - 7$	$5 - 6$	$5 - 10$	$7 - 3$
$1 - 6$	$10 - 5$	$9 - 10$	$8 - 3$	$1 - 3$
$7 - 9$	$1 - 2$	$10 - 0$	$3 - 5$	$6 - 9$
$4 - 8$	$6 - 1$	$0 - 9$	$9 - 4$	$7 - 8$
$8 - 4$	$0 - 4$	$2 - 6$	$6 - 9$	$9 - 5$

Page 24

1.

a. $\begin{array}{c} 1 + 3 = 4 \\ 4 - 3 = 1 \\ \text{or } 4 - 1 = 3 \end{array}$	b. $\begin{array}{c} 2 + 3 = 5 \\ 5 - 3 = 2 \\ \text{or } 5 - 2 = 3 \end{array}$
c. $\begin{array}{c} 4 + 5 = 9 \\ 9 - 4 = 5 \\ \text{or } 9 - 5 = 4 \end{array}$	d. $\begin{array}{c} 5 + 2 = 7 \\ 7 - 5 = 2 \\ \text{or } 7 - 2 = 5 \end{array}$

Page 25

1. (continued)

e. $\begin{array}{c} 2 + 4 = 6 \\ 6 - 2 = 4 \\ \text{or } 6 - 4 = 2 \end{array}$	f. $\begin{array}{c} 1 + 6 = 7 \\ 7 - 1 = 6 \\ \text{or } 7 - 6 = 1 \end{array}$
g. $\begin{array}{c} 4 + 5 = 9 \\ 9 - 4 = 5 \\ \text{or } 9 - 5 = 4 \end{array}$	h. $\begin{array}{c} 3 + 7 = 10 \\ 10 - 3 = 7 \\ \text{or } 10 - 7 = 3 \end{array}$
i. $\begin{array}{c} 3 + 3 = 6 \\ 6 - 3 = 3 \\ \text{or } 6 - 3 = 3 \end{array}$	j. $\begin{array}{c} 5 + 3 = 8 \\ 8 - 5 = 3 \\ \text{or } 8 - 3 = 5 \end{array}$
k. $\begin{array}{c} 2 + 6 = 8 \\ 8 - 2 = 6 \\ \text{or } 8 - 6 = 2 \end{array}$	l. $\begin{array}{c} 8 + 4 = 12 \\ 12 - 8 = 4 \\ \text{or } 12 - 4 = 8 \end{array}$

Page 26

2.

XXXXXX X a. $6 + 1 = 7$ $7 - 1 = 6$ $7 - 6 = 1$	XXXX XXX b. $4 + 3 = 7$ $7 - 4 = 3$ $7 - 3 = 4$
XXXXXX X c. $7 + 1 = 8$ $8 - 1 = 7$ $8 - 7 = 1$	XXXXX XXXX d. $5 + 4 = 9$ $9 - 5 = 4$ $9 - 4 = 5$
XX XXXXX e. $2 + 6 = 8$ $8 - 2 = 6$ $8 - 6 = 2$	XXX XXXXX f. $3 + 5 = 8$ $8 - 3 = 5$ $8 - 5 = 3$
X XXXXXXX g. $1 + 8 = 9$ $9 - 1 = 8$ $9 - 8 = 1$	XXXXX XXXXX h. $5 + 5 = 10$ $10 - 5 = 5$ $10 - 5 = 5$

Two Parts — One Total, pp. 27-29

There are five blue marbles and some white marbles in a bag. <u>Four are white.</u>	$5 + 4 = 9$ $9 - 5 = 4$

1. a. There are five blue flowers.

$5 + 5 = 10$
$10 - 5 = 5$

b. There are five girls.

$4 + 5 = 9$
$9 - 4 = 5$

Page 28

c. Two of the socks are black.

$8 + 2 = 10$
$10 - 8 = 2$

d. Six chairs were still standing upright on the lawn.

$2 + 6 = 8$
$8 - 2 = 6$

2. Answers will vary. For example:
   a. $8 - 2 = 6$. Gladys drew 8 circles and colored two of them yellow. The rest she colored red. There are six red circles.
   b. $7 - 3 = 4$. Together Joe and his sister had seven toys. Joe had three cars. Joe's sister had four dolls.

Page 29

3. a. $2 + 1 + 3 = 6$ b. $3 + 2 + 2 = 7$ c. $1 + 4 + 3 = 8$ d. $3 + 2 + 4 = 9$

4.

a. $3 + 2 + 3 = 8$	b. $1 + 5 + 4 = 10$

5. a. $2 + 2 + 6 = 10$ There were six yellow roses
   b. $1 + 2 + 4 = 7$ Four birds were brown.
   c. $2 + 2 + 5 = 9$ Five of her pencils are short.

Fact Families, pp. 30-33

Page 30

1. a. $1 + 5 = 6$, $5 + 1 = 6$, $6 - 1 = 5$, $6 - 5 = 1$ b. $3 + 5 = 8$, $5 + 3 = 8$, $8 - 3 = 5$, $8 - 5 = 3$
   c. $3 + 6 = 9$, $6 + 3 = 9$, $9 - 3 = 6$, $9 - 6 = 3$ d. $7 + 3 = 10$, $3 + 7 = 10$, $10 - 3 = 7$, $10 - 7 = 3$

Page 31

2. a. $0 + 7 = 7$, $7 + 0 = 7$, $7 - 0 = 7$, $7 - 7 = 0$ b. $1 + 6 = 7$, $6 + 1 = 7$, $7 - 1 = 6$, $7 - 6 = 1$
   c. $2 + 5 = 7$, $5 + 2 = 7$, $7 - 2 = 5$, $7 - 5 = 2$ d. $3 + 4 = 7$, $4 + 3 = 7$, $7 - 3 = 4$, $7 - 4 = 3$

3. a. Correct   b. Should be $\underline{5} - 4 = 1$ c. Should be $5 - \underline{3} = 2$
   d. Should be $\underline{5} - 4 = 1$ e. Should be $10 - 8 = \underline{2}$ f. Correct

Fact Families, cont.

Page 32

4.

a. Numbers: 5, 3, 2
  $2 + 3 = 5$
  $3 + 2 = 5$
  $5 - 3 = 2$
  $5 - 2 = 3$
	b. Numbers: 9, 4, 5
	  $4 + 5 = 9$
	  $5 + 4 = 9$
	  $9 - 4 = 5$
	  $9 - 5 = 4$

c. Numbers: 4, 0, 4
  $4 + 0 = 4$
  $0 + 4 = 4$
  $4 - 0 = 4$
  $4 - 4 = 0$
	d. Numbers: 10, 3, 7
	  $3 + 7 = 10$
	  $7 + 3 = 10$
	  $10 - 7 = 3$
	  $10 - 3 = 7$

e. Numbers: 10, 2, 8
  $2 + 8 = 10$
  $8 + 2 = 10$
  $10 - 8 = 2$
  $10 - 2 = 8$
	f. Numbers: 6, 0, 6
	  $6 + 0 = 6$
	  $0 + 6 = 6$
	  $6 - 0 = 6$
	  $6 - 6 = 0$

Page 33

5.

a. Numbers: 10, 5, 5
  $5 + 5 = 10$
  $5 + 5 = 10$
  $10 - 5 = 5$
  $10 - 5 = 5$
	b. Numbers: 9, 1, 8
	  $1 + 8 = 9$
	  $8 + 1 = 9$
	  $9 - 8 = 1$
	  $9 - 1 = 8$

c. Numbers: 6, 3, 3
  $3 + 3 = 6$
  $3 + 3 = 6$
  $6 - 3 = 3$
  $6 - 3 = 3$
	d. Numbers: 7, 1, 6
	  $1 + 6 = 7$
	  $6 + 1 = 7$
	  $7 - 6 = 1$
	  $7 - 1 = 6$

Puzzle Corner.

$9 - 4 = 5$	$5 - 1 = 4$	$1 + 4 = 5$
$5 - 2 = 3$	$7 - 6 = 1$	$1 + 2 = 3$
$8 - 7 = 1$	$10 - 5 = 5$	$3 + 7 = 10$

How Many More?, pp. 34-36

Page 34

1. a. 2 more, 2 fewer   b. 3 more, 3 fewer   c. 4 more, 4 fewer   d. 5 more, 5 fewer   e. 7 more, 7 fewer   f. 5 more, 5 fewer

Page 35

2. a. Jane has 9 marbles.   b. Mary has 6 marbles.   c. Eric has 2 marbles.   d. Jane has 2 marbles   e. Bill has 6 marbles.
   f. Liz has 5 marbles. g. Ed has 8 marbles.   h. Mary has 7 marbles.   i. Sue has 3 marbles.   j. Mary has 9 marbles.

Page 36

3. Check the student's work.

4. a. 2 more.   b. 3 years.   c. 2 years.   d. 5 fewer dolls.

"How Many More" Problems and Differences, pp. 37-40

1. a. $1 + 3 = 4$ b. $3 + 3 = 6$ c. $3 + 4 = 7$ d. $4 + 6 = 10$ e. $2 + 6 = 8$ f. $1 + 4 = 5$

2. a. $2 + 5 = 7$ b. $4 + 2 = 6$ c. $7 + 1 = 8$ d. $5 + 2 = 7$ e. $3 + 4 = 7$ f. $3 + 3 = 6$

3. a. 4 steps   b. 3 steps   c. 4 steps   d. 0 steps   e. 9 steps

4.

From	8	4	1	3	6	10	8	9
To	10	10	9	1	5	5	12	15
Difference	2	6	8	2	1	5	4	6

5.

a. from 3 to 5 __2__ steps $3 + 2 = 5$	b. from 1 to 5 __4__ steps $1 + 4 = 5$	c. from 2 to 7 __5__ steps $2 + 5 = 7$

6.

a. from 6 to 9 __3__ steps $6 + 3 = 9$	b. from 4 to 8 __4__ steps $4 + 4 = 8$	c. from 8 to 9 __1__ step $8 + 1 = 9$	d. from 2 to 6 __4__ steps $2 + 4 = 6$

7. a. $7 + 3 = 10$. Jill has __3__ more.   b. $4 + 3 = 7$.   Al has 3 more.
   c. $4 + 2 = 6$. Ann has two more.   d. $2 + 7 = 9$.   Hannah has 7 more.
   e. $10 + 1 = 11$. Britney has 1 more.   f. $5 + 5 = 10$. Don has 5 more.

8. a. $2 + 8 = 10$. There are 10 tapes. $2 + 6 = 8$ OR $8 - 2 = 6$. There are 6 tapes more on the shelf.
   b. $5 + 4 = 9$.   There are now 9 birds now in the oak tree.
      $5 + 4 = 9$ OR $9 - 5 = 4$. There are 4 more birds in the oak tree.
   c. $9 + 1 = 10$ OR $10 - 9 = 1$. Joe has 1 more car than Jason.
      $2 + 7 = 9$ OR $9 - 2 = 7$. Jason has 7 more cars than Brenda.

"How Many More" Problems and Subtraction, pp. 41-44

1.

a. $5 + 2 = 7$ $7 - 5 = 2$	b. $3 + 5 = 8$ $8 - 3 = 5$	c. $4 + 5 = 9$ $9 - 4 = 5$
d. $5 + 5 = 10$ $10 - 5 = 5$	e. $4 + 3 = 7$ $7 - 4 = 3$	f. $5 + 3 = 8$ $8 - 5 = 3$
g. $3 + 7 = 10$ $10 - 3 = 7$	h. $2 + 7 = 9$ $9 - 2 = 7$	i. $1 + 6 = 7$ $7 - 1 = 6$

"How Many More" Problems and Subtraction, cont.

Page 42

2.

a. $2 + 6 = 8$ $8 - 2 = 6$	b. $1 + 8 = 9$ $9 - 1 = 8$	c. $7 + 3 = 10$ $10 - 3 = 7$	d. $6 + 3 = 9$ $9 - 3 = 6$

3.

a. $1 + 6 = 7$ $7 - 1 = 6$	b. $2 + 7 = 9$ $9 - 2 = 7$	c. $1 + 9 = 10$ $10 - 1 = 9$
d. $5 + 3 = 8$ $8 - 3 = 5$	e. $8 + 2 = 10$ $10 - 2 = 8$	f. $6 + 3 = 9$ $9 - 3 = 6$

4. a. $2 + 6 = 8$. They ate 8 carrots.   b. $3 + 4 = 7$. The baby used 7 blocks.
   c. $5 + 2 = 7$. There are now 7 lambs.   d. $4 + 4 = 8$ or $8 - 4 = 4$. She needs 4 dollars.

Page 43

5. a. $8 - 6 = 2$, $6 + 2 = 8$ b. $10 - 9 = 1$, $9 + 1 = 10$ c. $9 - 7 = 2$, $7 + 2 = 9$
   d. $10 - 8 = 2$, $8 + 2 = 10$ e. $9 - 8 = 1$, $8 + 1 = 9$ f. $7 - 6 = 1$, $6 + 1 = 7$

6. a. $3 + 3 = 6$ or $6 - 3 = 3$. She needs 3 cucumbers.   b. $7 - 3 = 4$. Four ducks are left.
   c. $6 + 2 = 8$ or $8 - 6 = 2$. She needs two dollars.   d. $6 + 4 = 10$ or $10 - 6 = 4$. He has four pages to read.

7. a. Correct   b. Should be $8 - \underline{4} = 4$ c. Should be $\underline{6} - 4 = 2$ d. Correct   e. Correct   f. Should be $9 - \underline{7} = 2$

Page 44

8. a. 4 dolls   b. 6 teddy bears   c. 4 other toys

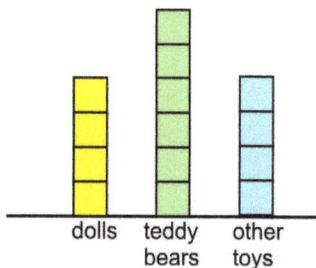

dolls   teddy bears   other toys

   d. 2 more   e. 2 more   f. 10 dolls and teddy bears in all

Subtracting More Than One Number, pp. 45-46

Page 45

1. a. 3, 1, 4   b. 5, 3, 2   c. 2, 2, 5

2. a. She has six cookies left. $10 - 2 - 2 = 6$ b. There are three birds left. $7 - 3 - 1 = 3$
   c. Now there are three cars. $8 - 3 - 2 = 3$ d. Jack has now $8. $5 + 2 + 1 = 8$

Page 46

3. a. 2, 1   b. 1, 2   c. 1, 3

4. a. 5, 4   b. 1, 0   c. 2, 1

5.

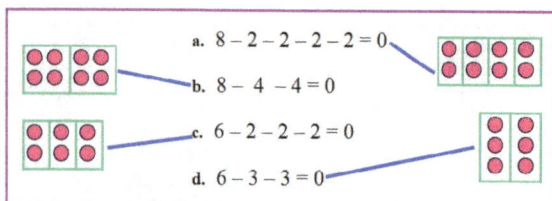

a. $8 - 2 - 2 - 2 - 2 = 0$
b. $8 - 4 - 4 = 0$
c. $6 - 2 - 2 - 2 = 0$
d. $6 - 3 - 3 = 0$

Puzzle corner. $9 - 3 - 2 - 1 = 3$ $10 - 1 - 2 - 1 = 6$ $8 - 4 - 1 - 2 = 1$

Page 47

1. $2 + 7 = 9, 7 + 2 = 9, 9 - 2 = 7, 9 - 7 = 2$

2. a. $10 - 4 = 6$ or $10 - 6 = 4$
   b. $5 + 4 = 9$; $9 - 5 = 4$

3. a. $8 - 2 = 6$. <u>Six of them are boys.</u>

3. b. $4 + 2 = 6$.

Susan's Kay's

c. $2 + 1 = 3$ robins. $5 - 3 = 2$.
   <u>Now there are two more sparrows.</u>

4. a. 1, 8, 7, 5   b. 3, 3, 1, 2   c. 10, 2, 0, 3   d. 6, 4, 9, 7

Addition and Subtraction Facts with 4 and 5, pp. 48-49

Page 48

	🦬🦬🦬 🦬🦬🦬	$5 + 0 = 5$ $5 - 5 = 0$ $0 + 5 = 5$ $5 - 0 = 5$
Facts with 5	🦬🦬🦬🦬 🦬	$4 + 1 = 5$ $5 - 4 = 1$ $1 + 4 = 5$ $5 - 1 = 4$
	🦬🦬🦬 🦬🦬	$3 + 2 = 5$ $5 - 3 = 2$ $2 + 3 = 5$ $5 - 2 = 3$

1. a. 1, 3, 4, 2   b. 3, 4, 1, 2   c. 5, 1, 3, 3   d. 4, 1, 4, 2

Page 49

2.

5 − 4	2 + 3	4 − 4	1 + 2	4 − 2	1 + 3
2 + 2	3 − 2	5 − 0	0 + 0	5 − 2	1 + 1
0 + 2	5 − 1	0 + 1	1 + 4	0 − 0	4 − 1

3.

$17 - 0 = 17$	$10 + 0 = 10$	$5 - 2 = 3$
$17 - 1 = 16$	$10 + 1 = 11$	$6 - 2 = 4$
$17 - 2 = 15$	$10 + 2 = 12$	$7 - 2 = 5$
$17 - 3 = 14$	$10 + 3 = 13$	$8 - 2 = 6$
$17 - 4 = 13$	$10 + 4 = 14$	$9 - 2 = 7$
$17 - 5 = 12$	$10 + 5 = 15$	$10 - 2 = 8$
$17 - 6 = 11$	$10 + 6 = 16$	$11 - 2 = 9$
$17 - 7 = 10$	$10 + 7 = 17$	$12 - 2 = 10$
$17 - 8 = 9$	$10 + 8 = 18$	$13 - 2 = 11$
$17 - 9 = 8$	$10 + 9 = 19$	$14 - 2 = 12$
$17 - 10 = 7$	$10 + 10 = 20$	$15 - 2 = 13$
$17 - 11 = 6$	$10 + 11 = 21$	$16 - 2 = 14$
$17 - 12 = 5$	$10 + 12 = 22$	$17 - 2 = 15$
etc.	etc.	etc.

Page 50

1.

6, 0, 6
$6 + 0 = 6$ $0 + 6 = 6$ $6 - 6 = 0$ $6 - 0 = 6$

5, 1, 6
$5 + 1 = 6$ $1 + 5 = 6$ $6 - 5 = 1$ $6 - 1 = 5$

4, 2, 6
$4 + 2 = 6$ $2 + 4 = 6$ $6 - 4 = 2$ $6 - 2 = 4$

3, 3, 6
$3 + 3 = 6$ $6 - 3 = 3$

2.

$0 + 6 = 6$	or	$6 + 0 = 6$
$1 + 5 = 6$	or	$5 + 1 = 6$
$2 + 4 = 6$	or	$4 + 2 = 6$
$3 + 3 = 6$		

3.

a. $\begin{array}{r} 6 \\ -5 \\ \hline 1 \end{array}$ b. $\begin{array}{r} 6 \\ -4 \\ \hline 2 \end{array}$ c. $\begin{array}{r} 6 \\ -6 \\ \hline 0 \end{array}$ d. $\begin{array}{r} 6 \\ -2 \\ \hline 4 \end{array}$ e. $\begin{array}{r} 6 \\ -1 \\ \hline 5 \end{array}$ f. $\begin{array}{r} 6 \\ -3 \\ \hline 3 \end{array}$

Page 51

5. a. 2, 3   b. 1, 6   c. 4, 5   d. 4, 1

6.

a. $2 + 3 = 5$ $5 - 2 = 3$	b. $1 + 5 = 6$ $6 - 1 = 5$	c. $4 + 1 = 5$ $5 - 4 = 1$
d. $3 + 5 = 8$ $8 - 3 = 5$	e. $5 + 5 = 10$ $10 - 5 = 5$	f. $2 + 5 = 7$ $7 - 2 = 5$

7.

a. $1 + 5 = 6$ $5 + 1 = 6$ $6 - 5 = 1$ $6 - 1 = 5$	b. $2 + 7 = 9$ $7 + 2 = 9$ $9 - 2 = 7$ $9 - 7 = 2$

Addition and Subtraction Facts with 6, cont.

Page 52

8. a. $4 + 3 = 7.$ <u>They have seven kittens in total.</u>
   $4 - 3 = 1$ OR $3 + 1 = 4.$ <u>The black cat has one more than the white one.</u>
   b. $10 - 2 = 8$ OR $2 + 8 = 10.$ <u>He has lost eight crayons.</u>
   c. $10 + 2 = 12.$ <u>Mother had 12 clothespins.</u> $10 - 2 = 8$ OR $8 + 2 = 10.$
   <u>Mother had eight more in the first container.</u>
   d. $8 - 2 = 6$ OR $2 + 6 = 8.$ <u>She needs six more eggs.</u>
   $2 + 3 = 5.$ <u>Jill and the neighbor have five eggs in total.</u>
   $2 + 3 = 5.$ <u>If she gets her neighbor's eggs, she will have five eggs.</u>
   $8 - 5 = 3$ OR $5 + 3 = 8.$ <u>She will need three more.</u>

Addition and Subtraction Facts with 7, pp. 53-54

Page 53

1.

7, 0, 7	6, 1, 7	5, 2, 7
$7 + 0 = 7$ $0 + 7 = 7$ $7 - 0 = 7$ $7 - 7 = 0$	$6 + 1 = 7$ $1 + 6 = 7$ $7 - 6 = 1$ $7 - 1 = 6$	$5 + 2 = 7$ $2 + 5 = 7$ $7 - 5 = 2$ $7 - 2 = 5$

4, 3, 7
$4 + 3 = 7$ $3 + 4 = 7$ $7 - 4 = 3$ $7 - 3 = 4$

2.

$0 + 7 = 7$	or	$7 + 0 = 7$
$1 + 6 = 7$	or	$6 + 1 = 7$
$2 + 5 = 7$	or	$5 + 2 = 7$
$3 + 4 = 7$	or	$4 + 3 = 7$

Page 54

4. a. $\begin{array}{r} 7 \\ -5 \\ \hline 2 \end{array}$ b. $\begin{array}{r} 7 \\ -4 \\ \hline 3 \end{array}$ c. $\begin{array}{r} 7 \\ -6 \\ \hline 1 \end{array}$ d. $\begin{array}{r} 7 \\ -2 \\ \hline 5 \end{array}$ e. $\begin{array}{r} 7 \\ -1 \\ \hline 6 \end{array}$ f. $\begin{array}{r} 7 \\ -3 \\ \hline 4 \end{array}$

5.

$5 + \underline{\hphantom{0}} = 7$	$7 - \underline{\hphantom{0}} = 4$	$7 - \underline{\hphantom{0}} = 6$
$7 - 0 = \underline{\hphantom{0}}$	$6 + \underline{\hphantom{0}} = 7$	$\underline{\hphantom{0}} + 4 = 7$
$7 - 3 = \underline{\hphantom{0}}$	$7 - 2 = \underline{\hphantom{0}}$	$7 - \underline{\hphantom{0}} = 7$
$7 - \underline{\hphantom{0}} = 1$	$0 + \underline{\hphantom{0}} = 7$	$7 - \underline{\hphantom{0}} = 2$

6. a. Jeremy has two more pencils than Luis. $6 - 4 = 2$ or $4 + \underline{\hphantom{0}} = 6.$
   In total, they have 10 pencils. $6 + 4 = 10$
   b. She found eight socks. $2 + 5 + 1 = 8.$

83

Addition and Subtraction Facts with 7, cont.

Page 54

Puzzle corner:

–	12	11	10	9	8	7	6	5	4	3
1	11	10	9	8	7	6	5	4	3	2
2	10	9	8	7	6	5	4	3	2	1

Addition and Subtraction Facts with 8, pp. 55-58

Page 55

1.

8, _0_, 8
$8 + 0 = 8$ $0 + 8 = 8$ $8 - 0 = 8$ $8 - 8 = 0$

_7_, 1, 8
$7 + 1 = 8$ $1 + 7 = 8$ $8 - 1 = 7$ $8 - 7 = 1$

6, 2, 8
$6 + 2 = 8$ $2 + 6 = 8$ $8 - 2 = 6$ $8 - 6 = 2$

5, 3, 8
$5 + 3 = 8$ $3 + 5 = 8$ $8 - 5 = 3$ $8 - 3 = 5$

4, 4, 8
$4 + 4 = 8$ $8 - 4 = 4$

Page 56

3.

$0 + \underline{8} = 8$ or $\underline{8} + 0 = 8$

$1 + \underline{7} = 8$ or $\underline{7} + 1 = 8$

$2 + \underline{6} = 8$ or $\underline{6} + 2 = 8$

$3 + \underline{5} = 8$ or $\underline{5} + 3 = 8$ $4 + \underline{4} = 8$

4. a. 5, 7, 6   b. 3, 7, 6   c. 4, 2, 1   d. 1, 8, 4

5.

a. $5 - 2$	4	b. $7 - 4$	5	c. $8 - 1$	7	d. $6 - 3$	2
↓	↓	↓	↓	↓	↓	↓	↓
3 < 4		3 < 5		7 = 7		3 > 2	

Addition and Subtraction Facts with 8, cont.

Page 57

6.

a. $5-2$ $4-2$ ↓ ↓ **3** > **2**	b. $8-1$ $7-1$ ↓ ↓ **7** > **6**	c. $8-6$ $8-5$ ↓ ↓ **2** < **3**
d. $6+2$ $7+2$ ↓ ↓ **8** < **9**	e. $7-1$ $7-2$ ↓ ↓ **6** > **5**	f. $4+4$ $7-5$ ↓ ↓ **8** > **2**
g. $1-1$ $3-2$ ↓ ↓ **0** < **1**	h. $3+10$ 10 ↓ ↓ **13** > **10**	i. 7 $4+2$ ↓ ↓ **7** > **6**
j. $8-1$ $4-2$ ↓ ↓ **7** > **2**	k. $7-2$ $6-1$ ↓ ↓ **5** = **5**	l. $9-0$ $7+2$ ↓ ↓ **9** = **9**

Page 58

7.

$6+2=8$	$8-4=4$	$3+5=8$
$8-0=8$	$8-5=3$	$8-2=6$
$8-3=5$	$7+1=8$	$4+4=8$
$7+1=8$	$8-2=6$	$8-0=8$
$4+4=8$	$0+8=8$	$8-1=7$

8. a. $7-4=3$. Bill has three more cars than Ed.
   $10-4=6$. Jack has six more cars than Ed.
   $10-7=3$. Jack has three more cars than Bill.
   b. $5+3=8$. Both things together cost $8, so she cannot buy both. She has $7.
   $7+1=8$. She would need $1 more.

Addition and Subtraction Facts with 9, pp. 59-61

Page 59

1.

9, 0 , 9	8, 1, 9	7, 2, 9
★★★★★ ★★★★	★★★★ ★ ★★★★	★★★★ ★ ★★★ ★
$9+0=9$ $0+9=9$ $9-0=9$ $9-9=0$	$8+1=9$ $1+8=9$ $9-1=8$ $9-8=1$	$7+2=9$ $2+7=9$ $9-7=2$ $9-2=7$

Page 59

1. (continued)

6, 3, 9	5, 4, 9
★★★ ★★ ★★★ ★	★★★ ★★ ★★ ★★
$6 + 3 = 9$ $3 + 6 = 9$ $9 - 6 = 3$ $9 - 3 = 6$	$5 + 4 = 9$ $4 + 5 = 9$ $9 - 5 = 4$ $9 - 4 = 5$

Page 60

3.

$0 + \underline{9} = 9$ or $\underline{9} + 0 = 9$
$1 + \underline{8} = 9$ or $\underline{8} + 1 = 9$
$2 + \underline{7} = 9$ or $\underline{7} + 2 = 9$
$3 + \underline{6} = 9$ or $\underline{6} + 3 = 9$
$4 + \underline{5} = 9$ or $\underline{5} + 4 = 9$

4.   a. 4, 6, 3, 1   b. 7, 8, 2, 1   c. 8, 6, 4, 2   d. 8, 9, 7, 5

5.

$7 + \underline{2} = 9$	$0 + \underline{9} = 9$	$\underline{1} + 8 = 9$
$9 - 3 = \underline{6}$	$9 - \underline{6} = 3$	$9 - \underline{3} = 6$
$9 - 1 = \underline{8}$	$5 + \underline{4} = 9$	$\underline{7} + 2 = 9$
$9 - \underline{0} = 9$	$9 - 2 = \underline{7}$	$9 - \underline{5} = 4$
$4 + \underline{5} = 9$	$9 - \underline{8} = 1$	$\underline{9} + 0 = 9$

Page 61

6. a. $8 \boxed{>} 10 - 3$ b. $9 \boxed{<} 9 + 3$ c. $8 - 6 \boxed{<} 6 + 3$

    d. $6 + 2 \boxed{<} 8 + 2$ e. $10 - 1 \boxed{<} 10$ f. $8 - 4 \boxed{>} 8 - 5$

    g. $5 - 2 \boxed{>} 4 - 2$ h. $8 + 0 \boxed{=} 8 - 0$ i. $9 - 1 \boxed{<} 9 + 1$

7. a. $\begin{array}{r} 9 \\ -5 \\ \hline 4 \end{array}$ b. $\begin{array}{r} 9 \\ -4 \\ \hline 5 \end{array}$ c. $\begin{array}{r} 9 \\ -6 \\ \hline 3 \end{array}$ d. $\begin{array}{r} 8 \\ -2 \\ \hline 6 \end{array}$ e. $\begin{array}{r} 9 \\ -2 \\ \hline 7 \end{array}$ f. $\begin{array}{r} 8 \\ -3 \\ \hline 5 \end{array}$

Addition and Subtraction Facts with 9, cont.

Page 61

8.

$9-3$	$4+6$	$9-0$	$4+1$	$8-1$
$2+5$	$9-5$	$4+4$	$4-2$	$5+1$
$9-2$	$3+7$	$10-2$	$10+0$	$7-1$
$4+2$	$7-3$	$6+3$	$3-1$	$3+3$
$6-0$	$1+1$	$8-0$	$3+2$	$10-4$
$3+4$	$8-3$	$2+7$	$7-6$	$7+0$
$1+6$	$2+8$	$10-1$	$2+2$	$7-0$

($9-3$, $2+5$, $9-2$, $4+2$, $6-0$, $3+4$, $1+6$, $8-1$, $5+1$, $7-1$, $3+3$, $10-4$, $7+0$, $7-0$ are colored blue, $9-0$, $4+4$, $10-2$, $6+3$, $8-0$, $2+7$, $10-1$ are colored red, and the rest are colored yellow.)

Addition and Subtraction Facts with 10, pp. 62-65

Page 62

1.

10, 0, 10
$10 + 0 = 10$
$0 + 10 = 10$
$10 - 10 = 0$
$10 - 0 = 10$

9, 1, 10
$9 + 1 = 10$
$1 + 9 = 10$
$10 - 9 = 1$
$10 - 1 = 9$

8, 2, 10
$8 + 2 = 10$
$2 + 8 = 10$
$10 - 2 = 8$
$10 - 8 = 2$

7, 3, 10
$7 + 3 = 10$
$3 + 7 = 10$
$10 - 7 = 3$
$10 - 3 = 7$

6, 4, 10
$6 + 4 = 10$
$4 + 6 = 10$
$10 - 6 = 4$
$10 - 4 = 6$

5, 5, 10
$5 + 5 = 10$
$10 - 5 = 5$

Page 63

3.

$0 + \underline{10} = 10$	or	$\underline{10} + 0 = 10$
$1 + \underline{9} = 10$	or	$\underline{9} + 1 = 10$
$2 + \underline{8} = 10$	or	$\underline{8} + 2 = 10$
$3 + \underline{7} = 10$	or	$\underline{7} + 3 = 10$
$4 + \underline{6} = 10$	or	$\underline{6} + 4 = 10$
$5 + \underline{5} = 10$		

4. a. 7, 4, 8   b. 10, 3, 1   c. 3, 8, 6

Page 63

5.

6 + 4 = 10	0 + 10 = 10	5 + 5 = 10
10 − 5 = 5	8 + 2 = 10	10 − 4 = 6
10 − 1 = 9	10 − 7 = 3	8 + 2 = 10
10 − 2 = 8	5 + 5 = 10	10 − 3 = 7
10 − 0 = 10	10 − 4 = 6	0 + 10 = 10
7 + 3 = 10	9 + 1 = 10	10 − 9 = 1

Page 64

6. a. 2 + 3 = 5. Elisa has five coins. Sarah still has more coins. 6 − 5 = 1.
   The difference is one coin.
   b. 6 + 1 = 7. Dad got a total of seven boxes. Then he gave one away. 7 − 1 = 6.
   So he ended up with six boxes.
   c. 6 + 3 = 9. We have a total of nine dollars, so yes, we can buy the meal. 9 − 8 = 1.
   We will have one dollar left over.

7. a. 5, 3, 8   b. 7, 6, 9   c. 4, 3, 1

Page 65

8.

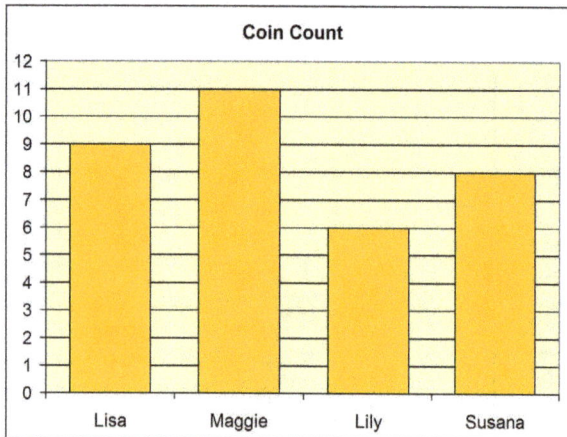

9. Answers will vary; check the student's questions and answers. For example:
   How many more coins does Lisa have than Maggie?

Puzzle Corner.
Use "guess and check" to find the answers.

a. The square is 6, the triangle is 4. 6 + 4 = 10, 6 − 4 = 2
b. The square is 8, the triangle is 2. 8 + 2 = 10, 8 − 2 = 6
c. The square is 5, and the triangle is 5. 5 + 5 = 10, 5 − 5 = 0

Review - Facts with 6, 7 and 8, pp. 66-67

Page 66

1.

a.	b.	c.	d.
$0 + 8 = 8$	$3 + 4 = 7$	$6 - 4 = 2$	$7 - 5 = 2$
$3 + 5 = 8$	$5 + 2 = 7$	$6 - 1 = 5$	$8 - 5 = 3$
$2 + 6 = 8$	$1 + 6 = 7$	$6 - 3 = 3$	$6 - 5 = 1$
$6 + 2 = 8$	$6 + 1 = 7$	$6 - 2 = 4$	$8 - 4 = 4$
$5 + 3 = 8$	$2 + 5 = 7$	$6 - 5 = 1$	$7 - 3 = 4$

2.

a. $8 - 2 \boxed{} 7 - 3$ $\downarrow \qquad \downarrow$ $6 > 4$	b. $10 - 7 \boxed{} 9 - 6$ $\downarrow \qquad \downarrow$ $3 = 3$	c. $7 - 6 \boxed{} 4 - 2$ $\downarrow \qquad \downarrow$ $1 < 2$
d. $\quad 4 + 2 > 9 - 8$	e. $\quad 10 - 4 > 7 - 4$	f. $\quad 3 + 4 > 7 - 1$

3. a. <u>Luisa had 5 more counters.</u> $9 - 4 = 5$ or $4 + 5 = 9$.
   b. <u>Luisa had three more counters.</u> $8 - 5 = 3$ or $5 + 3 = 8$.

Page 67

4.

$6 - 5 = 1$	$7 - 5 = 2$	$8 - 3 = 5$
$2 + 5 = 7$	$4 + 2 = 6$	$5 + 2 = 7$
$8 - 5 = 3$	$6 - 1 = 5$	$1 + 5 = 6$
$6 + 2 = 8$	$5 + 3 = 8$	$8 - 6 = 2$
$6 - 4 = 2$	$8 - 2 = 6$	$2 + 4 = 6$

5.

$3 + 4 = 7$	$1 + 6 = 7$	$8 - 1 = 7$
$6 - 3 = 3$	$8 - 7 = 1$	$1 + 6 = 7$
$7 + 1 = 8$	$3 + 3 = 6$	$3 + 3 = 6$
$8 - 4 = 4$	$4 + 3 = 7$	$8 - 4 = 4$
$7 - 1 = 6$	$8 - 4 = 4$	$7 - 3 = 4$

Puzzle Corner.
Answers will vary.
Here is one
possibility.

2	$+$	6	$=$	8
$-$		$-$		
2	$+$	4	$=$	6
$\parallel$		$\parallel$		
0		2		

3	$+$	4	$=$	7
$+$		$-$		
5	$-$	1	$=$	4
$\parallel$		$\parallel$		
8		3		

89

Page 68

1.

a.	b.	c.	d.
$4 + \underline{5} = 9$	$5 + \underline{5} = 10$	$10 - \underline{9} = 1$	$9 - \underline{7} = 2$
$1 + \underline{8} = 9$	$2 + \underline{8} = 10$	$10 - \underline{3} = 7$	$9 - \underline{3} = 6$
$6 + \underline{3} = 9$	$3 + \underline{7} = 10$	$10 - \underline{5} = 5$	$9 - \underline{1} = 8$
$2 + \underline{7} = 9$	$4 + \underline{6} = 10$	$10 - \underline{2} = 8$	$9 - \underline{4} = 5$

2.

a. $2 + 3 + 3 = 8$

b. $3 + 2 + 2 = 7$

c. $1 + 2 + 2 = 5$

d. $3 + 4 + 2 = 9$

e. $3 + 3 + 3 = 9$

f. $2 + 2 + 2 = 6$

3. a. Correct   b. Should be $10 - \underline{4} = 6$ c. Should be $9 - 4 = \underline{5}$ d. Correct   e. Should be $7 - \underline{4} = 3$ f. Correct

Page 69

4.

$9 - 2 = 7$	$7 + 2 = 9$	$9 - 4 = 5$
$9 - 3 = 6$	$8 + 1 = 9$	$9 - 6 = 3$
$9 - 1 = 8$	$4 + 5 = 9$	$9 - 7 = 2$
$9 - 5 = 4$	$3 + 6 = 9$	$1 + 8 = 9$

5. a. Answers may vary. Please check the student's work. In part (a) one number 9 will be left unpaired.
   In part (b) a number 6 will be left unpaired.

6.

$10 - 2 = 8$	$1 + 9 = 10$	$9 + 1 = 10$
$10 - 5 = 5$	$4 + 6 = 10$	$10 - 5 = 5$
$10 - 9 = 1$	$5 + 5 = 10$	$10 - 4 = 6$
$10 - 3 = 7$	$2 + 8 = 10$	$7 + 3 = 10$
$10 - 6 = 4$	$3 + 7 = 10$	$10 - 2 = 8$

Page 70

7. a. $7 - 2 = 5$. <u>Ken has five more than Millie.</u>   b. $3 + 4 + 3 = 10$. <u>Mike has ten cars.</u>

   c. $4 + 4 = 8$. <u>There were eight birds.</u> $8 - 5 = 3$. <u>Later, there were three birds.</u>

   d. $4 + 6 = 10$ or $10 - 4 = 6$. <u>Six crayons are missing.</u>   e. $10 - 2 = 8$. <u>There are eight pieces left.</u>

More from math MAMMOTH

Math Mammoth has a variety of resources to fit your needs. All are available as economical downloads, and most also as printed copies.

- **Math Mammoth Light Blue Series**
  A complete curriculum for grades 1-7. Each grade level includes two student worktexts (A and B), which contain all the instruction and exercises all in the same book, answer keys, tests, cumulative reviews, and a worksheet maker. International (all metric), Canadian, and South African versions are also available.

  https://www.MathMammoth.com/complete-curriculum

  https://www.MathMammoth.com/international/international

  https://www.MathMammoth.com/canada/

  https://www.MathMammoth.com/south_africa/

- **Math Mammoth Skills Review Workbooks**
  These workbooks are intended to be used alongside the Light Blue series full curriculum, and they provide additional review to the topics studied in the main curriculum, in a spiral manner.
  https://www.MathMammoth.com/skills_review_workbooks/

- **Math Mammoth Blue Series**
  Blue Series books are topical worktexts for grades 1-7, containing both instruction and exercises. The topics cover all elementary mathematics from 1st through 7th grade. These books are not tied to grade levels, and are thus great for filling in gaps.
  https://www.MathMammoth.com/blue-series

- **Make It Real Learning**
  These activity workbooks concentrate on answering the question, "Where is math used in real life?" The series includes various workbooks for grades 3-12.
  https://www.MathMammoth.com/worksheets/mirl/

- **Review Workbooks**
  Workbooks for grades 1-7 that provide a comprehensive review of one grade level of math—for example, for review during school break or summer vacation.
  https://www.MathMammoth.com/review_workbooks/

Free gift!

- Receive over 350 free sample pages and worksheets from my books, plus other freebies:
  https://www.MathMammoth.com/worksheets/free

Lastly...

- Inspire4 is an inspirational website for the whole family I've been privileged to help with:
  https://www.inspire4.com

www.ingramcontent.com/pod-product-compliance
Lightning Source LLC
Chambersburg PA
CBHW051229200326
41519CB00025B/7300